T0140262

Studies in Computational Intelligence

Volume 566

Series editor

Janusz Kacprzyk, Polish Academy of Sciences, Warsaw, Poland
e-mail: kacprzyk@ibspan.waw.pl

About this Series

The series "Studies in Computational Intelligence" (SCI) publishes new developments and advances in the various areas of computational intelligence—quickly and with a high quality. The intent is to cover the theory, applications, and design methods of computational intelligence, as embedded in the fields of engineering, computer science, physics and life sciences, as well as the methodologies behind them. The series contains monographs, lecture notes and edited volumes in computational intelligence spanning the areas of neural networks, connectionist systems, genetic algorithms, evolutionary computation, artificial intelligence, cellular automata, self-organizing systems, soft computing, fuzzy systems, and hybrid intelligent systems. Of particular value to both the contributors and the readership are the short publication timeframe and the world-wide distribution, which enable both wide and rapid dissemination of research output.

More information about this series at http://www.springer.com/series/7092

Roger Lee

Editor

Computer and Information Science

 Springer

Editor
Roger Lee
Software Engineering and Information
 Technology Institute
Central Michigan University
Mt. Pleasant, MI
USA

ISSN 1860-949X   ISSN 1860-9503   (electronic)
ISBN 978-3-319-36172-7   ISBN 978-3-319-10509-3   (eBook)
DOI 10.1007/978-3-319-10509-3

Springer Cham Heidelberg New York Dordrecht London

Printed on acid-free paper

Springer is part of Springer Science+Business Media (www.springer.com)

Foreword

The purpose of the 13th IEEE/ACIS International Conference on Computer and Information Science (ICIS 2014) held during June 4–6, 2014 in Taiyuan, China, was to gather researchers, scientists, engineers, industry practitioners, and students to discuss, encourage, and exchange new ideas, research results, and experiences on all aspects of Applied Computers and Information Technology, and to discuss the practical challenges encountered along the way and the solutions adopted to solve them. The conference organizers selected the best 14 papers from those papers accepted for presentation at the conference in order to publish them in this volume. The papers were chosen based on review scores submitted by members of the program committee and underwent further rigorous rounds of review.

In "A New Method of Breakpoint Connection for Human Skeleton Image," Xiaoping Li, Degui Zhao, Yongliang Hu, Ye Song, Na Fu, Qiongxin Liu present a new breakpoint algorithm based on layer and partition of the neighborhood. The algorithm scans skeleton images line by line. The number of other skeleton points is calculated for each skeleton point in its 8-neighborhood, then judges whether this skeleton point is a breakpoint or not according to the number of the above obtained and the distribution of other skeleton points in its 8-neighborhood.

In "Insult Detection in Social Network Comments Using Possibilistic Based Fusion Approach," Mohamed Maher Ben Ismail and Ouiem Bchir propose a novel approach to automatically detect verbal offense in social network comments. It relies on a local approach that adapts the fusion method to different regions of the feature space in order to classify comments from social networks as insult or not. The proposed algorithm is formulated mathematically through the minimization of some objective function. It combines context identification and multi-algorithm fusion criteria into a joint objective function.

In "What Information in Software Historical Repositories Do We Need to Support Software Maintenance Tasks? An Approach Based on Topic Model," Xiaobing Sun, Bin Li, Yun Li, and Ying Chen propose a preprocess to facilitate selection of related SHR to support various software maintenance tasks. The preprocess uses the topic model to extract the related information from Software

Historical Repositories (SHR) to help support software maintenance, thus improving the effectiveness of traditional SHR-based technique. Empirical results show the effectiveness of this approach.

In "Evaluation Framework for the Dependability of Ubiquitous Learning Environment," Manel BenSassi and Mona Laroussi introduce a proposed framework to evaluate ubiquitous learning that treats the issue of considering contextual dimensions from a technological point of view. This framework is considered in the research interested in developing ubiquitous learning environments based on wireless and sensor technologies. Finally, they detail how they exploit this framework to evaluate a realistic case study.

In "Improving Content Recommendation in Social Streams via Interest Model," Junjie Zhang and Yongmei Lei implement three recommendation engines based on Sina Micro-blog and deploy them online to gather feedback from real users. Experimental results show that this method can recommend interesting information to users and improve the precision and stability of personalized information recommendation by 30 %.

In "Performance Evaluation of Unsupervised Learning Techniques for Intrusion Detection in Mobile Ad Hoc Networks," Binh Hy Dang and Wei Li demonstrate a research effort to evaluate the effectiveness and efficiency of different unsupervised detection techniques. Different types of experiments were conducted, with each experiment involving different parameters such as number of nodes, speed, pause time, among others. The proposed evaluation methodology provides empirical evidence on the choice of unsupervised learning algorithms, and could shed light on the future development of novel intrusion detection techniques for MANETs.

In "Live Migration Performance Modelling for Virtual Machines with Resizable Memory," Cho-Chin Lin, Zong-De Jian and Shyi-Tsong Wu present a general model for live migration. An effective strategy for optimizing the service downtime under this model is suggested. The performance of live migration is evaluated for virtual machines with resizable memory.

In "A Heuristic Algorithm for Workflow-Based Job Scheduling in Decentralized Distributed Systems with Heterogeneous Resources," Nasi Tantitharanukul, Juggapong Natwichai, and Pruet Boonma address the problem of job scheduling, so-called workflow-based job scheduling, in decentralized distributed systems with heterogeneous resources. As this problem is proven to be an NP-complete problem, an efficient heuristic algorithm to address this problem is proposed. The algorithm is based on an observation that the heterogeneity of resources can affect the execution time of scheduling. They compare the effectiveness and efficiency of the proposed algorithm with a baseline algorithm.

In "Novel Data Integrity Verification Schemes in Cloud Storage," Thanh Cuong Nguyen, Wenfeng Shen, Zhaokai Luo, Zhou Lei, and Weimin Xu propose two alternative schemes, called DIV-I and DIV-II, to verify that cloud data has not been illegally modified. Compared to S-PDP introduced by Ateniese et al., both DIV-I and DIV-II use less time to generate tags and verify. In addition, the proposed schemes fully support dynamic operations as well as public verification.

In "Generation of Assurance Cases For Medical Devices," Chung-Ling Lin and Wuwei Shen take the medical systems industry into account to illustrate how an assurance case can be generated when a software process is employed. In particular, we consider the Generic Insulin infusion Pump (GIIP) to show how an assurance case can be produced via a popular software development process, called Rational Unified Process (RUP).

In "A Survey on the Categories of Service Value/Quality/Satisfactory Factors," Yucong Duan, Nanjangud C. Narendra, Bo Hu, Donghong Li, Wenlong Feng, Wencai Du, and Junxing Lu work toward a solution for the missing factors in Service modeling standardization. They use the constructive process to classify the factors into more than 20 higher level categories with explanations on the process.

In "Effective Domain Modeling for Mobile Business AHMS (Adaptive Human Management Systems) Requirements," Haeng-Kon Kim and Roger Lee suggest a method that systematically defines, analyzes, and designs a domain to enhance reusability effectively in Mobile Business Domain Modeling (MBDM) in Adaptive Human Management Systems (AHMS) requirements phase.

In "A New Modified Elman Neural Network with Stable Learning Algorithms for Identification of Nonlinear Systems," Fatemeh Nejadmorad Moghanloo, Alireza Yazdizadeh, and Amir Pouresmael Janbaz Fomani propose a new dynamic neural network structure, based on the Elman Neural Network (ENN), for identification of nonlinear. Encouraging simulation results reveal that the idea of using the proposed structure for identification of nonlinear systems is feasible and very appealing.

In "A Simple Model for Evaluating Medical Treatment Options," Irosh Fernando, Frans Henskens, Masoud Talebian, and Martin Cohen introduce a model that is intuitive to clinicians for evaluating medication treatment options, and therefore has the advantage of engaging clinicians actively in a collaborative development of clinical Decision Support Systems (DSS).

It is our sincere hope that this volume provides stimulation and inspiration, and that it will be used as a foundation for works to come.

June 2014

Wenai Song
Simon Xu
Lichao Chen

Contents

Contributors

Ouiem Bchir College of Computer and Information Sciences, King Saud University, Riyadh, Kingdom of Saudi Arabia

Mohamed Maher Ben Ismail College of Computer and Information Sciences, King Saud University, Riyadh, Kingdom of Saudi Arabia

Henda BenGhezela RIADI GDL, Ensi, University of Manouba, Manouba, Tunisia

Manel BenSassi RIADI GDL, Ensi, University of Manouba, Manouba, Tunisia

Pruet Boonma Data Engineering and Network Technology Laboratory, Faculty of Engineering, Department of Computer Engineering, Chiang Mai University, Chiang Mai, Thailand

Ying Chen School of Information Engineering, Yangzhou University, Yangzhou, China

Martin Cohen The Mater Hospital, Hunter New England Area Health Service, Waratah, NSW, Australia

Binh Hy Dang Graduate School of Computer and Information Sciences, Nova Southeastern University, Fort Lauderdale, FL, USA

Wencai Du College of Information Science and Technology, Hainan University, Haikou, China

Yucong Duan College of Information Science and Technology, Hainan University, Haikou, China

Wenlong Feng College of Information Science and Technology, Hainan University, Haikou, China

Irosh Fernando School of Electrical Engineering and Computer Science, University of Newcastle, Callaghan, NSW, Australia

Na Fu Department of Computer Science and Technology, Beijing Institute of Technology, Beijing, China

Frans Henskens School of Electrical Engineering and Computer Science, University of Newcastle, Callaghan, NSW, Australia

Bo Hu Kingdee International Software Group China, Hong Kong, China

Yongliang Hu Department of Modern Distance Education, Beijing Institute of Technology, Beijing, China

Zong-De Jian Department of Electronic Engineering, National Ilan University, Yilan, Taiwan

Haeng-Kon Kim School of Information Technology, Catholic University of Daegu, Hayang, South Korea

Mona Laroussi RIADI GDL, Ensi, University of Manouba, Manouba, Tunisia

Roger Y. Lee Department of Computer Science, Central Michigan University, Michigan, USA

Yongmei Lei Computer Engineering and Science, Shanghai University, Shanghai, China

Zhou Lei School of Computer Engineering and Science, Shanghai University, Shanghai, China

Bin Li School of Information Engineering, Yangzhou University, Yangzhou, China

Donghong Li School of Statistics and Mathematics, Central University of Finance and Economics, Beijing, China

Wei Li Graduate School of Computer and Information Sciences, Nova Southeastern University, Fort Lauderdale, FL, USA

Xiaoping Li Department of Modern Distance Education, Beijing Institute of Technology, Beijing, China

Yun Li School of Information Engineering, Yangzhou University, Yangzhou, China

Cho-Chin Lin Department of Electronic Engineering, National Ilan University, Yilan, Taiwan

Chung-Ling Lin Department of Computer Science, Western Michigan University, Kalamazoo, MI, USA

Qiongxin Liu Department of Computer Science and Technology, Beijing Institute of Technology, Beijing, China

Junxing Lu College of Information Science and Technology, Hainan University, Haikou, China

Zhaokai Luo School of Computer Engineering and Science, Shanghai University, Shanghai, China

Nanjangud C. Narendra Cognizant Technology Solutions, Bangalore, India

Juggapong Natwichai Data Engineering and Network Technology Laboratory, Faculty of Engineering, Department of Computer Engineering, Chiang Mai University, Chiang Mai, Thailand

Fatemeh Nejadmorad Moghanloo Department of Electrical Engineering, Abbaspour College of Technology, Shahid Beheshti University, Tehran, Iran

Thanh Cuong Nguyen School of Computer Engineering and Science, Shanghai University, Shanghai, China

Amir Pouresmael Janbaz Fomani Department of Electrical Engineering, Abbaspour College of Technology, Shahid Beheshti University, Tehran, Iran

Wenfeng Shen School of Computer Engineering and Science, Shanghai University, Shanghai, China

Wuwei Shen Department of Computer Science, Western Michigan University, Kalamazoo, MI, USA

Ye Song Department of Computer Science and Technology, Beijing Institute of Technology, Beijing, China

Xiaobing Sun School of Information Engineering, Yangzhou University, Yangzhou, China

Masoud Talebian School of Mathematical and Physical Sciences, University of Newcastle, Callaghan, NSW, Australia

Nasi Tantitharanukul Data Engineering and Network Technology Laboratory, Faculty of Engineering, Department of Computer Engineering, Chiang Mai University, Chiang Mai, Thailand

Shyi-Tsong Wu Department of Electronic Engineering, National Ilan University, Yilan, Taiwan

Weimin Xu School of Computer Engineering and Science, Shanghai University, Shanghai, China

Alireza Yazdizadeh Department of Electrical Engineering, Abbaspour College of Technology, Shahid Beheshti University, Tehran, Iran

Junjie Zhang Computer Engineering and Science, Shanghai University, Shanghai, China

Degui Zhao Department of Computer Science and Technology, Beijing Institute of Technology, Beijing, China

A New Method of Breakpoint Connection for Human Skeleton Image

Xiaoping Li, Degui Zhao, Yongliang Hu, Ye Song, Na Fu and Qiongxin Liu

Abstract There are many discontinuous skeleton points in human skeleton images, which make human skeleton behavior analysis be a difficult problem. Our paper presents a new breakpoint algorithm based on layer and partition of the neighborhood. We scan skeleton images line by line. The number of other skeleton points is calculated for each skeleton point in its 8-neighborhood. Then we judge whether this skeleton point is a breakpoint or not according to the number of the above-obtained and the distribution of other skeleton points in its 8-neighborhood. If it is, we will find available connection skeleton points which can connect the breakpoint. Finally, we find out the points that need to be updated to complete the breakpoint connection progress in accordance with linear equations established by the breakpoint and available connection points. Through of theory analysis and experiment verification, our method has good effect on connecting breakpoints in skeleton images and the shapes of skeleton images are undeformed. In addition to the human skeleton images, this method can also be used for other objects skeleton images on the breakpoints connection.

Keywords Breakpoint connection · Neighborhood · Skeleton image

1 Introduction

The behavioral status judgement of video object is not only applied in application fields such as video retrieval [1], intelligent national defense and public security [2], but also provided high value in a wide range of applications under extreme conditions

X. Li · Y. Hu
Department of Modern Distance Education,
Beijing Institute of Technology, Beijing 100081, China
e-mail: lxpmx@x263.net

D. Zhao (✉) · Y. Song · N. Fu · Q. Liu
Department of Computer Science and Technology,
Beijing Institute of Technology, Beijing 100081, China
e-mail: zdgwangyi@163.com

© Springer International Publishing Switzerland 2015
R. Lee (ed.), *Computer and Information Science*, Studies in Computational
Intelligence 566, DOI 10.1007/978-3-319-10509-3_1

and the environment which is difficult to observe, like the status judgement of extreme sports, the real-time analysis of astronauts' extravehicular behavioral status [3] and the locking of the military targets. However, there are so many types of monitoring video objects that have big discrepancies and different characteristics among them. So there is a need to choose a symbolic object as a representative for research. The human behavior analysis [4] has become the focus issue and representative research in the field of behavior recognition [5] because there are many features such as complexity, diversity and covering a wide range in human behavior. One of the important methods in human behavior analysis is to extract human body contour, skeleton the human body and at last use vector processing for the human skeleton model. The technology of modeling human skeleton can reduce amount of information needed to express the human body and retain the human behavioral characteristics at the same time. Existing skeleton modeling algorithm can be divided into three categories: the central axis conversion method [6], iterative morphological method [7] and ZS refinement [8]. Among these three categories of skeleton modeling algorithm, the first one is proposed earliest and the most well known. Both the Zhang-Suen thinning algorithm [9] proposed by Zhang and Fu and the Holt [10] thinning algorithm are belong to the second category. The third method mentioned above may lead to the location of nodes offset because it only deals with one side at a time. Due to the fact that some broken curves in the skeleton image [11] which is obtained by skeleton modeling algorithm, we can not locate the human skeleton key points and do human behavior analysis subsequently. That is to say connection of breakpoints plays a vital role in human behavior analysis.

Until now, many researchers devote themselves to this topic and propose several kinds of methods to try to solve the problem. The first is Method of the minimum value [12]: this method takes each breakpoint with the minimum distance or direction difference as its connection. It will make errors especially when the points are far away from each other. The second is Method of mathematical morphology [13]: this method repairs breakpoints by the basic operators like erosion, dilation, opening and closing. It is suitable for the simple situations but for the complex situations it works not so well. The last one is Method of graph search [14]: it builds a graph based on breakpoints and makes the best connections through graph search under the necessary physical rules. This method has very high time complexity.

Different with the methods of the above-mentioned, we present a method based on layer and partition of the neighborhood for discontinuous skeleton in this paper.

2 Image Preprocessing

For a human body movement image, we need to get the image of the human body contour using background subtraction method. For the contour image denoising, we use mathematical morphology such as opening and closing to preserve details and smooth non-impulsive noise. Then Otsu method [15] is applied to binarize the contour image. And next, central axis algorithm [16] and Holt algorithm are utilized for image thinning. After the preprocessing, the skeleton image is composed of discrete fracture curves. So connection operation needs to be done subsequently for the existing breakpoints and extract the human contour completely.

3 Basic Elements

In this paper, we divide all the breakpoints to two types by the number of other skeleton points in the breakpoint's 8-neighborhood. The basic elements are explained below.

3.1 Neighborhood Layer

Each pixel has 8 neighboring pixels except boundary points in an image. They are called 8-neighborhood, which is named as the first neighborhood layer in our paper. And the second neighborhood layer consists of the next 16 nearest pixels which are called 16-neighborhood shown in Fig. 1. Like this, the third neighborhood layer includes 24 pixels and the N neighborhood layer has pixels of N multiplied by 8. The distributions of neighborhood layers are shown in Fig. 2. The same color points belong to a same neighborhood layer.

P1	P2	P3	P4	P5
P16				P6
P15		P		P7
P14				P8
P13	P12	P11	P10	P9

Fig. 1 16-Neighborhood of p point

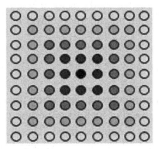

Fig. 2 Layers of neighborhood

Fig. 3 *Upper left* and *lower*
right neighborhood of the
second neighborhood layer

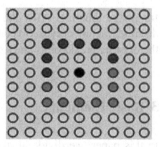

3.2 Upper Left Neighborhood and Lower Right Neighborhood

In our paper, a neighborhood is divided into the upper left and lower right neighborhood according to a point is detected or not in the N neighborhood layer. If a point had been detected before, it belongs to the upper left neighborhood; otherwise, it is classified into the lower right neighborhood. The upper left and lower right neighborhood of the second neighborhood layer are shown in Fig. 3. The upper left neighborhood is consist of red points. So the green points belong to the lower right neighborhood.

3.3 Breakpoint Type

We classify all the breakpoints to two types according to the number of other skeleton points in the 8-neighborhood:

3.3.1 Isolated Point

When a skeleton point has no other skeleton points in its 8-neighborhood, this skeleton point is considered as an isolated point. Then if we try to find its available skeleton points for connection, we firstly need to search the upper left and lower right neighborhood of the second neighborhood layer.

3.3.2 Common Breakpoint

We call a breakpoint as a common breakpoint when the number of skeleton points is 1 in the upper left neighborhood of its 8-neighborhood and that of the lower right neighborhood equals to 0 in its 8-neighborhood. This time we only find the lower right neighborhood for available skeleton points in the second neighborhood layer.

3.4 Available Connection Point

In breakpoint's neighborhood, some skeleton points are called available connection point when they are used for the breakpoint connection. Isolated point need two available connection points at least and one at least for common breakpoint. We search available connection points from the lower neighborhood to the high neighborhood. If there is an available connection point in a lower neighborhood, firstly we connect the available connection point and the current skeleton breakpoint, then search for the other skeleton breakpoints.

4 Breakpoint Connection Function

Cartesian coordinate systems are established, in which each one of available connection points is the coordinate origin. The coordinate of the available connection point is (CpW, CpH) and that of skeleton breakpoint is (BpW, BpH). Then we can get ΔW and ΔH.

$$\Delta W = BpW - CpW$$
$$\Delta H = CpH - BpH$$

4.1 $\Delta W = 0$

When ΔW equals to 0, the connection between the available point and the breakpoint is in a special circumstance. Some points need to be updated to skeleton points as shown in the Fig. 4. In our paper, the black point in the coordinate origin is an available connection point and another one is the breakpoint. The red color points will be updated to skeleton points.

Fig. 4 Schematic diagram of
updating a skeleton point
when $\Delta W = 0$

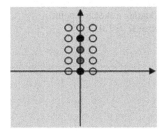

Fig. 5 Schematic diagram of
updating a skeleton point
when K = 1/3

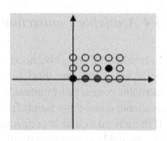

4.2 ΔW ≠ 0

When ΔW is not equal to 0, a liner equation is established by an available connection point and a breakpoint. According to the value of its slope K, we can get the coordinates of points that need to be updated to skeleton points. All of the following 6 charts illustrate connections of all the conditions between the breakpoint and an available connection point in the lower right neighborhood. The connection in the upper left neighborhood is similar.

4.2.1 0.0 < K < 1.0

Take K = 1/3 for example, the coordinates of points that need to be updated to skeleton points are (CpW + i, CpH − ⌊i*K⌋) and i is an integer from 1 to (BpW - CpW) (Fig. 5).

4.2.2 K = 1.0

When K equals 1.0, the coordinates of points that need to be updated are (CpW + i, CpH − i) and i is an integer from 1 to (BpW - CpW) (Fig. 6).

Fig. 6 Schematic diagram of
updating a skeleton point
when K = 1.0

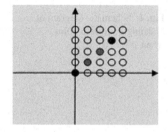

Fig. 7 Schematic diagram of updating a skeleton point when $K = 3/2$

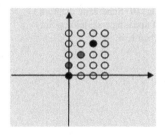

Fig. 8 Schematic diagram of updating a skeleton point when $K = -2/3$

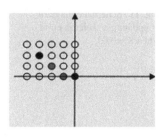

Fig. 9 Schematic diagram of updating a skeleton point when $K = -1.0$

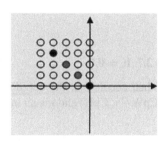

4.2.3 K > 1.0

Take $K = 3/2$ for example, the coordinates of points that need to be updated are $(CpW + \lfloor i/K \rfloor, CpH - i)$ and i is an integer from 1 to (CpH - BpH) (Fig. 7).

4.2.4 −1.0 < K < 0.0

Take $K = -2/3$ for example, the coordinates of points that need to be updated are $(CpW - i, CpH + \lceil i*K \rceil)$ and i is an integer from 1 to (CpW - BpW) (Fig. 8).

4.2.5 K = −1.0

When K equals −1.0, the coordinates of points that need to be updated are $(CpW - i, CpH - i)$ and i is an integer from 1 to (CpW - BpW) (Fig. 9).

4.2.6 K < −1.0

Take $K = -3/2$ for example, the coordinates of skeleton points that need to be updated are $(CpW + \lceil i/K \rceil, CpH - i)$ and i is an integer from 1 to (CpH - BpH) (Fig. 10).

Fig. 10 Schematic diagram
of updating a skeleton point
when K = −3/2

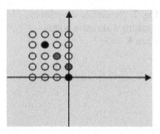

Fig. 11 Schematic diagram
of updating a skeleton point
when K = 0.0

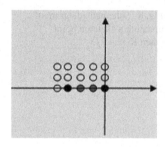

4.2.7 K = 0.0

When equals 0.0, the coordinates of skeleton points that need to be updated are
(CpW − i, CpH) and i is an integer from 1 to (CpW - BpW) (Fig. 11).

5 Procedure of Breakpoint Connection

5.1 Search Skeleton Breakpoint

The skeleton image is a binary image, so we use the following method to search
breakpoints in it. The image is scanned line by line. When a pixel is a skeleton point,
we calculate the number of other skeleton points in its 8-neighborhood and judge the
type of the breakpoint.

5.2 Search Available Connection Point

There are some differences between common breakpoints and isolated points when
we try to search available connection points. If the point is a common breakpoint,
we only need to search the lower right neighborhood of the second neighborhood
layer. But when it is an isolated point, all points of the second neighborhood layer
will be detected. Moreover, all the available connection points will be classified to
the points of the upper left or the lower right neighborhood; note the numbers and
coordinates separately.

5.3 Connect Breakpoint

If available connection points are found in breakpoint's second neighborhood layer, then we can establish a Cartesian coordinate system at every one of available connection points and calculate the slopes of linear equations which are established by the breakpoint and each one of the available connection points. Then we use Breakpoint connection function to update corresponding pixels. But if not, we need to increase the number of the neighborhood layer and keep on searching. Once we find an available connection point, stop searching and connect with the skeleton breakpoint. Record the breakpoint information if we can not find an available connection point within the range of the threshold Dn of neighborhood layer and keep on searching for other skeleton breakpoints.

For these breakpoints which are recorded, we increase the threshold Dn of neighborhood layer and find available connection points again. The times of increasing Dn depends on the value of scanning times which can be set in advance. For example, when the value of scanning times is 3 and the initial value of the threshold Dn is 2, that means we will only scan the second neighborhood layer the first time for all breakpoints. The second and third neighborhood layer will be scanned at the second time for the rest of breakpoints and at the third time we will scan the second, third and fourth neighborhood for the rest of the second time.

6 Experimental Results

In our paper, the algorithm has been tested with real images as shown in the below figures. In each group, the original image is at the left side and the right side image is the connection result from our method. We scan the skeleton image under different thresholds of neighborhood. The information of image in seven groups are shown in Fig. 12. The initial value of neighborhood layer is two and increase one in every times of scanning. We use Windows 7 system to run our program. The system parameters are i5–3470, 3.20 GHz and 4.00 GB RAM. The type of our system is 32-bit operating system (Fig. 13).

Group A:

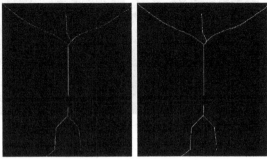

Group B:

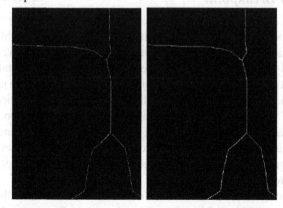

Group C:

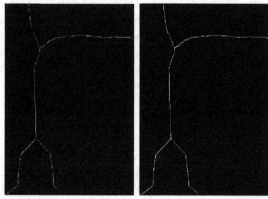

Group D:

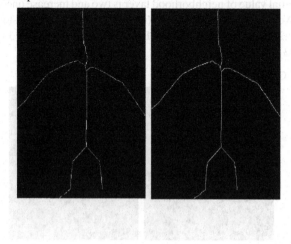

Group E:

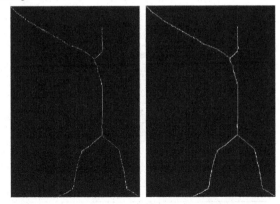

Group F:

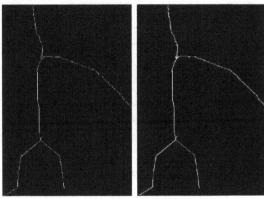

Group G:

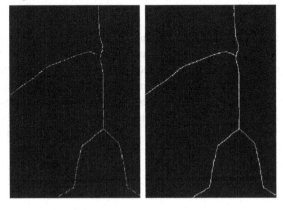

Name of Image	Width of Image	Height of Image	Size of Image	Format of Image
A	438	484	4.79KB	.gif
B	338	461	3.75KB	.gif
C	322	447	5.46KB	.gif
D	365	476	4.71KB	.gif
E	300	501	4.25KB	.gif
F	247	446	3.35KB	.gif
G	304	452	4.17KB	.gif

Fig. 12 Information of seven groups images

	Before connection of breakpoints		After connection of breakpoints								
	Number of common breakpoints	Number of isolated points	Number of common breakpoints under the threshold D_n					Number of isolated points under the threshold D_n			
			D_n	3	4	5	6	D_n	3	4	5	6
Group A	100	61		11	3	1	1		1	1	1	0
Group B	88	40		9	4	0	0		0	0	0	0
Group C	91	47		10	6	4	2		0	0	0	0
Group D	118	44		7	6	3	1		2	0	0	0
Group E	102	75		9	3	3	1		1	0	0	0
Group F	92	53		11	4	4	3		0	0	0	0
Group G	110	66		11	5	2	2		2	1	1	1

Fig. 13 Contrast of the breakpoints number before and after connections

7 Conclusion

Since there are many breakpoints in human skeleton image and this makes the human behavior analysis cannot go on. So in our paper, a novel algorithm for breakpoint connection is proposed based on layer and partition of the neighborhood. The proposed method has been tested by experiments on human skeleton images. And the whole program works well on those images. The original width of skeleton images is guaranteed because of the single-pixel connection. In addition to the human skeleton images, the proposed method also has a good effect on breakpoint connection of other objects skeleton images.

References

1. Alvarez-Alvarez, A., Trivino, G., Cord, O.: Body posture recognition by means of a genetic fuzzy finite state machine. In: IEEE 5th International Workshop, Genetic and evolutionary fuzzy systems (GEFS), vol. 4, pp. 60–65 (2011)

2. Schnell, T., Keller, M., Poolman, P.: A quality of training effectiveness assessment (QTEA): a neurophysiologically based method enhance flight training. Digital Avionics Systems Conference, vol. 10, pp. 4.D.6-1–4.D.6-13 (2008)
3. Hoinville, T., Naceri, A.: Performances of experienced and novice sportball players in heading virtual spinning soccer balls. In: Virtual Reality Conference (VR) vol. 3, pp. 83–86 (2011)
4. Ahad, M.A.R., Tan, J., Kim, H., Ishikawa, S.: Action recognition by employing combined directional motion history and energy images. In: IEEE Computer Society Conference on Computer Vision and Pattern Recognition Workshops (CVPRW), vol. 6, pp. 73–78 (2010)
5. Rosales, R., Sclaroff, S.: Specialized mappings and the estimation of human body pose from a single image. In: Workshop on Human Motion, vol. 3, pp. 19–24 (2000)
6. Lohou, C., Bertrand, G.: A 3D 6-subiteration curve thinning algorithm based on beta-simple points. Discrete Appl. Math. **151**, 1–3 (2005)
7. Fei, X., Guili, X., Yuehua, C.: An improved thinning algorithm for human body recognition. In: IEEE International Workshop on Imaging System and Techniques, pp. 411–415 (2009)
8. Yang, F., Shundong, Z.: Effective mixed fingerprint image thinning algorithm. In: Proceeding of Information Technology and Environmental System Sciences, pp. 116–122 (2008)
9. Basak, J., Pal, N.R., Patel, P.S.: Thinning in binary and gray images: a connectionist approach. J. Inst. Electron. Telecommun. Eng. **4**(42), 305–313 (1996)
10. Perrot, R.H., Holt, C., Clint, M., Stewart, A.: A parallel processing algorithm for thinning. Digitized Pictures Lect. Notes Comput. Sci. **237**, 183–189 (1986)
11. Hilaire, X., Tombre, K.: Robust and accurate vectorization of line drawings. IEEE Trans. Pattern Anal. Mach. Intell. **28**(6), 890–904 (2006)
12. Lu, T., Tai, C.L., Yang, H., Cai, S.: A novel knowledge-based system for interpreting complex engineering drawings: theory, representation, and implementation. IEEE Trans. Pattern Anal. Mach. Intell. **31**(8), 1444–1457 (2009)
13. Tuia, D., Pacifici, F., Kanevski, M.: Classification of very high spatial resolution imagery using mathematical morphology and support vector machines. IEEE Trans. Geosci. Remote Sens. **47**(11), 3866–3879 (2009)
14. Fernandez, J.A., Gonzalez, J.: Multihierarchical graph search. IEEE Trans. Pattern Anal. Mach. Intell. **24**(1), 103–113 (2002)
15. Zhang, G.Y., Liu, G.Z., Zhu, H., Qiu, B.: Ore image thresholding using bi-neighborhood otsu's approach. Electron. Lett. **46**(25), 1666–1668 (2010)
16. Lohou, C., Bertrand, G.: 3D 6-subiteration curve thinning algorithm based on beta-simple points. Discrete Appl. Math. **151**, 1–3 (2005)

Insult Detection in Social Network Comments Using Possibilistic Based Fusion Approach

Mohamed Maher Ben Ismail and Ouiem Bchir

Abstract This paper aims to propose a novel approach to automatically detect verbal offense in social network comments. It relies on a local approach that adapts the fusion method to different regions of the feature space in order to classify comments from social networks as insult or not. The proposed algorithm is formulated mathematically through the minimization of some objective function. It combines context identification and multi-algorithm fusion criteria into a joint objective function. This optimization is intended to produce contexts as compact clusters in subspaces of the high-dimensional feature space via possibilistic unsupervised learning and feature weighting. Our initial experiments have indicated that the proposed fusion approach outperforms individual classifiers and the global fusion method. Also, in order to validate the obtained results, we compared the performance of the proposed approach with related fusion methods.

Keywords Supervised learning · Fusion · Social networks · Insult detection

1 Introduction

The widespread of smart devices and broadband internet connections yield an exponential growth of social networks. These networks are hosted and managed by very big companies which are employing thousands of people, and investing millions of dollars in order to improve their services, features, and performance. Also, millions of users gathered within these virtual societies, and formed several communities sharing the same interest and skills. Thus, blogs and social networks have become very active spaces where people express, comment, and share their opinions. However, the cultural heterogeneity of some users yields some misunderstanding of

M.M. Ben Ismail (✉) · O. Bchir
College of Computer and Information Sciences, King Saud University,
Riyadh, Kingdom of Saudi Arabia
e-mail: mbenismail@ksu.edu.sa

O. Bchir
e-mail: obchirl@ksu.edu.sa

© Springer International Publishing Switzerland 2015
R. Lee (ed.), *Computer and Information Science*, Studies in Computational
Intelligence 566, DOI 10.1007/978-3-319-10509-3_2

each other comment meanings. In fact, a user may consider someone else comment inappropriate even though it was not meant to offense him. Moreover, sometimes verbal offenses, insults and other defamatory language shared by certain users cause hurt feelings, especially when they are addressed to "conservative" people. A natural solution to overcome this problem was to appoint human moderators monitoring online conversation. However, this solution can be expensive and labor intensive task for the moderator. Moreover, comments can be very frequent, which makes the process not efficient enough. The earliest efforts in this area were directed towards matching comments with a vocabulary of "prohibited" words. In other words, if the comment contains one or more keywords from the banned word list, then the comment is denied. These efforts posed the problem of insult detection as a string matching problem. In [1], the authors outlined a system which relies on a static dictionary and some patterns based on socio-linguistic. However, the obtained results proved that the approach suffers from high false positive rates and low coverage. The authors in [2] proposed an approach which consists in differentiating between insult and factive statements by parsing the sentences and using the semantic rules. The main drawback of this approach is its inability to discriminate efficiently between insults directed to non-participant and participant of conversation. The system proposed in [3] relies on a dictionary abusing language dictionary which is coupled with a bag-of words features. However, these state-of-the-arts show two main drawbacks. The first one consists of the use of seed words and naive matching approach. The second drawback is that these solutions are not able to distinguish between insults directed towards people participating in blog/forum conversation and non-participants such as celebrities, public figures etc. In other words, comments which contain racial slurs or profanity may not necessarily be insulting to other users. Using machine learning techniques was a natural alternative to automatically detect verbal offenses in social network comments. More specifically, the problem has been perceived as text classification problem.

During the last decade, several researches was proposed to overcome the challenge of automatic insult detection in social network comments. In [4], topical features and lexicon features and various machine learning techniques have been used in order to detect offensive tweets. In [5], the authors proposed an approach that exploits linguistic regularities in profane language via statistical topic modeling on comments corpus. Another flame detection software was proposed in [3]. This tool applies models from multi-level classifiers, boosted by a dictionary of Insulting and Abusing Language. One can notice that these state-of-the-art solutions rely on efficient supervised learning algorithms [6]. These algorithms learn models from known examples (labeled comments) and use them in order to automatically classify new samples (unlabeled comments).

Supervised learning algorithms have been extensively applied to several challenges in real world applications. For instance, in [7], the authors outlined how classification is used to perform opinion mining. The authors in [8–10] proposed several approaches in order to either automatically classify emails using the subject or detect junk email [11]. Usually, it is admitted that there is no one best way to solve the challenges and it may be useless to argue which type of classification technique is best [12]. Therefore, many approaches to combine the outputs of several classifiers were proposed in order to enhance the effectiveness of standard single classifier systems. Classifier fusion or ensemble classifier has become a very active research field, and promising results have been obtained with several applications [13]. Namely, fusion has been applied to pattern recognition, including character recognition [14], speech recognition [15], and text categorization [16], and have outperformed single classifier systems both theoretically and experimentally. Motivated by the classifiers' complementary characteristics, ensemble of classifiers can outperform individual algorithms by exploiting the advantages of the individual classifiers and limiting the effect of their disadvantages. Nonetheless, a necessary and sufficient conditions for a fusion classifier to be more accurate than single classifier are diversity and accuracy [17]. Classifier fusion methods rely on an effective combination of the classifiers outputs. This process considers all experts competitive and equally trained on the whole feature space. For unlabeled point, single experts are launched simultaneously. Then, the obtained outputs are combined in a way to take a group decision. The classifier combinations methods can be grouped based on how they assign weights to the single experts. Namely, global methods assign an average degree of worthiness over the feature space to each expert. On the other hand, local methods formulate the classifiers' worthiness with respect to different feature subspaces. These data-dependent weights, when learned properly more accurate classifiers. In [18], the authors outlined a method where they estimate the accuracy of each expert in local regions of the feature space neighboring an unlabeled test point. Then, the most accurate classifier in that specific local region is used for the final decision. However, the need to estimate the accuracy for each test sample makes the approach time-consuming. In [19], the clustering-and-selection method was proposed. Basically, this method selects statistically the best classifier. First, the training samples are clustered to form the decision regions. Then, the classifier that performed the best in terms of accuracy on this local region is chosen. However, the method was not generic enough to consider more than one classifiers for one region. The authors of [20] extended the clustering-and-selection approach, and exploited the class labels. In other words, they divided the training set into correctly and incorrectly classified samples. Then, they categorized them in order to form a partition of the feature space. For testing, they pick the most effective classifier based its accuracy in the vicinity of the input point in order to make the final decision. Thus, each classifier should maintain its own partition. This makes the decision process computationally expensive. Lately, in [21, 22] the authors presented a local fusion technique which partitions the feature space into homogeneous regions based on their features, and adopts the obtained feature space structure when performing the fusion. On the other hand, the fusion component assigns an aggregation weight to each detector in each context based on its relative performance within the context. However, the adopted fuzzy approach makes the fusion stage sensitive to outliers. In fact, outliers may affect the obtained partition and reduce the accuracy of the final decision.

To overcome this limitation, we propose a possibilistic based optimization to partition the feature space and the fusion of the classifiers. The partitioning of the feature space is based on the standard sum of within cluster distances. However, for complex classification problems, the data is usually noisy which yields inaccurate partitions of the feature space. To alleviate this drawback, we propose a possibilistic based local approach that adapts the fusion method to different regions of the feature space. The aggregation weights are then estimated by the fusion component to each detector. These weights assignment is based on the relative performance of each detector within the context. Categorizing the input samples into regions during the training phase is a main requirement of the fusion component. Then, this approach appoints an expert for each region. These experts represent the best classifiers for the corresponding region.

2 Fusion Based on Possibilistic Context Extraction

Let N training observations with desired output $T = \{t_j | j = 1, \ldots N\}$. These outputs were obtained using K classifiers. Each classifier k extracts its own feature set $X_k = \{x_j^k | j = 1, \ldots N\}$ and generates confidence values, $Y^k = \{y_{kj} | j = 1, \ldots N\}$. The K feature sets are then concatenated to generate one global descriptor, $\chi = \bigcup_{k=1}^{K} \chi^k = \{x_j = [x_j^1, \ldots, x_j^K | j = 1, \ldots, N]\}$. The original Context Extraction for Local Fusion algorithm [9] minimizes

$$J = \sum_{j=1}^{N} \sum_{i=1}^{C} u_{ji}^m \sum_{s=1}^{K} v_{ik}^q d_{ijk}^2 + \sum_{j=1}^{N} \sum_{i=1}^{C} \beta_i u_{ji}^m \left(\sum_{k=1}^{K} \omega_{ik} y_{kj} - t_j \right)^2, \quad (1)$$

subject to $\sum_{i=1}^{C} u_{ji} = 1 \, \forall j$, $u_{ji} \in [0, 1] \, \forall i, j$, $\sum_{k=1}^{K} v_{ik} = 1 \, \forall i$, $v_{ik} \in [0, 1] \, \forall i, k$, and $\sum_{k=1}^{K} \omega_{ik} = 1 \, \forall i$.

The first term in (1) corresponds to the objective function of the Fuzzy C-Means (FCM) algorithm [23]. It is intended to categorize the N points into C clusters centered in c_i. Each data point x_j will be assigned to all clusters with fuzzy membership degrees. When a partition of C compact clusters with minimum sum of intra-cluster distances is discovered this FCM term is minimized. The second term in (1) attempts to learn cluster-dependent aggregation weights of the K algorithm outputs. ω_{ik} is the aggregation weight assigned to classifier k within cluster i. This term is minimized when the aggregated partial output values match the desired output. When both terms are combined and β is chosen properly, the algorithm seeks to partition the data into compact and homogeneous clusters while learning optimal aggregation weights for each algorithm within each cluster.

For real world classification problems, multiple sources of information and multiple classifiers for each source may be needed to obtain satisfactory results. In this case, the resulting feature space can be noisy and high dimensional. This complicates the clustering task, and the true partition of the data cannot be generated. This is due to the influence of the noisy points on the obtained clusters. To alleviate this drawback, we propose a possibilistic version of the algorithm. The proposed algorithm generates possibilistic memberships in order to represent the degree of typicality of each data point within every category, and reduce the influence of noise points on the learning process. We extend the objective function (1), and formulate the context extraction for local fusion using the following objective function

$$J = \sum_{j=1}^{N} \sum_{i=1}^{C} u_{ji}^m \sum_{k=1}^{K} v_{ik}^q d_{ijk}^2 + \sum_{j=1}^{N} \sum_{i=1}^{C} \beta_i u_{ji}^m \left(\sum_{k=1}^{K} \omega_{ik} y_{kj} - t_j \right)^2$$

$$+ \sum_{i=1}^{C} \eta_i \sum_{j=1}^{N} (1 - u_{ji})^m, \tag{2}$$

In (2), u_{ji} represents the possibilistic membership of $\mathbf{X_j}$ in cluster i. The $M \times N$ matrix, $U = [u_{ji}]$ is called a possibilistic partition if it satisfies:

$$\begin{cases} u_{ji} \in [0, 1], & \forall j \\ 0 < \sum_{i=1}^{C} u_{ji} < N \; \forall i, j \end{cases} \tag{3}$$

On the other hand the $M \times d$ matrix of feature subset weight, $V = [v_{ik}]$ satisfies

$$\begin{cases} v_{ik} \in [0, 1] \; \forall i, k \\ \sum_{k=1}^{K} v_{ik} = 1 \; \forall i \end{cases} \tag{4}$$

In (2), $m \in [1, \infty)$ is called the fuzzier, and η_i are positive constants that controls the importance of the second term with respect to the first one. This term is minimized when u_{ji} are close to 1, thus, avoiding the trivial solution of the first term (where $u_{ji} = 0$). Note that $\sum_{i=1}^{C} u_{ji}$ is not constrained to sum to 1. In fact, points that are not representative of any cluster will have $\sum_{i=1}^{C} u_{ji}$ close to zero and will be considered as noise. This constraint relaxation overcomes the disadvantage of the constrained fuzzy membership approach which is the high sensitivity to noise and outliers. The parameter η_i is related to the resolution parameter in the potential function and the deterministic annealing approaches. It is also related to the idea of "scale" in robust statistics. In any case, the value of 0.7 determines the distance at which the membership becomes 0.5. The value of η_i determines the "zone of influence" of a point. A point $\mathbf{X_j}$ will have little influence on the estimates of the model parameters of a cluster if $\sum_{k=1}^{K} v_{ik}^q (d_{ijk})^2$ is large when compared with η_i. On the other hand, the "fuzzier" m determines the rate of decay of the membership value. When $m = 1$, the memberships are crisp. When $m \rightarrow \infty$, the membership function does not decay to zero at all. In this possibilistic approach, increasing values of m represent increased possibility of all points in the data set completely belonging to a given cluster.

Setting the gradient of J with respect to u_{ji} to zero yields the following necessary condition to update the possibilistic membership degrees [24]:

$$u_{ji} = \left[1 - \left(\frac{D_{ij}^2}{\eta_j} \right)^{\frac{1}{m-1}} \right]^{-1}. \tag{5}$$

where $D_{ij} = \sum_{k=1}^{K} v_{ik}^q d_{ijk}^2 + \beta \sum_{k=1}^{K} v_{ik}^q \left(\sum_{l=1}^{K} \omega_{il} y_{lj} - t_j \right)^2$. D_{ij} represents the total cost when considering point x_j in cluster i. As it can be seen, this cost depends on the distance between point x_j and the cluster's centroid c_i, and the deviation of the combined algorithms' decision from the desired output (weighted by β). More specifically, points to be assigned to the same cluster: (i) are close to each other in the feature space, and (ii) their confidence values could be combined linearly with the same coefficients to match the desired output.

Minimizing J with respect to the feature weights yields

$$v_{ik} = \sum_{l=1}^{K} \left[(D_{ik}^2/D_{il})^{\frac{1}{q-1}} \right] \tag{6}$$

where $D_{il} = \sum_{j=1}^{N} u_{ij}^m d_{ijl}^2$.

Minimization of J with respect to the prototype parameters, and the aggregation weights yields

$$c_{jk} = \frac{\sum_{j=1}^{N} u_{ij}^m \mathbf{X_{jk}}}{\sum_{j=1}^{N} u_{ij}^m}. \tag{7}$$

and

$$w_{ik} = \frac{\sum_{j=1}^{N} u_{ij}^m y_{kj} \left(t_j - \sum_{\substack{l=1 \\ l \neq k}}^{K} \omega_{il} y_{lj} \right) - \zeta_i}{\sum_{j=1}^{N} u_{ij}^m y_{kj}^2}. \tag{8}$$

where ζ_i is a Lagrange multiplier that assures that the constraint in (3) is satisfied, and is defined as

$$\zeta_i = \frac{\sum_{l=1}^{K} \frac{\sum_{j=1}^{N} u_{ij}^m y_{lj} \left(t_j - \sum_{k=1}^{K} \omega_{ik} y_{kj} \right)}{\sum_{j=1}^{N} u_{ij}^m y_{lj}^2}}{\sum_{l=1}^{K} \frac{1}{\sum_{j=1}^{N} u_{ij}^m y_{lj}^2}}. \tag{9}$$

The behavior of this algorithm depends on the value of β. Over estimating it yields the multi-algorithm fusion criteria to be dominant which results in non-compact clusters. On the other hand, a small value of β reduces the influence of the multi-algorithm fusion criteria and categorizes the data based mainly on the distances in the feature space.

The obtained algorithm is an iterative algorithm that starts with an initial partition and alternates between the update equations of u_{ji}, v_{ik}, and c_{ik}. It is summarized below.

Algorithm 1 Fusion algorithm based on Possibilistic Context Extraction

> **Begin**
> *Fix the number of clusters* C;
> *Fix* m, q *and* β.
> *Initialize the centers and the possibilistic* M *partition matrix* U;
> *Initialize the relevance weights to* $1/K$;
> **Repeat**
>   *Compute* d_{ijk}^2, *for* $1 \leq i \leq C$ *and* $1 \leq j \leq N$ *and* $1 \leq k \leq K$;
>   *Update the relevance weights* v_{ik} *using equation (6)*;
>   *Compute* D_{ij}^2
>   *Update the partition matrix* U *using equation (5)*;
>   *Update the partition matrix* W *using equation (8)*;
>   *Update the centers using equation (7)*;
> **Until** *(centers stabilize)*
> **End**

3 Experiments

A range of experiments were performed to asses the strengths and weaknesses of the proposed approach. We used the KAGGLE data [25]. This collection of social commentary consists of two subsets. The first one represents the training set with 3948 comments. The second subset is the testing collection, and it consists of 2235 comments.

First, we preprocessed the comments collection in order to discard some encoding parts that may affect the results, gather similar words with stemming and discard the less frequent words. For instance, a raw comments looks like "\ \ xc \ \xa0If you take out the fags and booze...". After preprocessing it, we obtain "If you take out the fags and booze...". Also, we deleted words starting with "@". Then, substituted words like "u" to "you", and "da" to "the" etc.

For feature extraction, we used standard TFiDF [26] technique in order to map comments x_j into a compact representation of its content. Thus, each comment x_j was represented using one 800-dimensional feature vector $x_j = \langle w_{1,j}, \ldots, w_{|\tau|,j} \rangle$. Where τ is the vocabulary of words that occur at least once in at least one comment, and $0 < w_{k,j} < 1$ represents how much the kth word contributes to the semantics of comment x_j.

Due to the space limitation, the effect of these parameters cannot be illustrated in this work. To adapt this data to our application, we assume that we have 3 sets of features and that we have one classifier for each set. These sets are extracted as subsets from the original features. Specifically, The first subset includes features from 1 to 400. The second one includes features from 200 to 600, and the third subset includes features 400–800. For each set, we use a simple K-NN classifier to generate confidence values. We classify the training data using the 3 K-NN classifiers with their appropriate feature subsets. Then, we use the proposed local fusion to partition the training data into 3 clusters. For each cluster, the algorithm learns the optimal aggregation weights. The testing phase starts by classifying the test point using the

Table 1 Confusion matrices obtained using three single classifiers, the Method in [9], and the proposed method, respectively

		Predicted as not insult	Predicted as insult
Classifier 1			
	Not insult	1449	505
	Insult	538	155
Classifier 2			
	Not insult	1352	602
	Insult	545	148
Classifier 3			
	Not insult	1448	506
	Insult	535	158
Method in [9]			
	Not insult	1320	591
	Insult	495	198
Proposed method			
	Not insult	1309	645
	Insult	430	290

four classifiers and generating the corresponding partial confidence values. Then, we assign it to the closest cluster. Finally, the final decision is obtained by combining the four partial confidence values using the aggregation weights of the closest cluster.

Table 1 shows the confusion matrix obtained using the three single classifiers, the Method in [9], and the proposed method. One can notice that the proposed method outperforms the other approaches in terms of Specificity. More specifically, our approach detected about 50 % more insult comments than single classifiers. On the other hand, our approach classifies less accurately non-insult comments which yields lower sensitivity value. Despite this discrepancy between sensitivity and specificity, we consider these results promising because for the problem of automatic detection of insults in social network comets, we assume that the True Negative predictions are not equally relevant to True Positive ones. In other words, we do not consider misclassifying an insult comment as serious as misclassifying a non-insult one. Moreover, since the testing data contains 720 insulting comments only out of 2674 comments, the accuracy cannot be an appropriate performance measure for this application. In Table 1, we show the aggregation weights learned by the proposed algorithms. As it can be seen, the relative performances of the individual classifiers varies significantly from one cluster to another. For instance for cluster 1, classifier 3 outperforms the other classifiers. Consequently, this classifier is considered the most reliable one for this cluster and is assigned the highest aggregation weight as shown in Table 2. Similarly, for cluster 3, classifier 2 is assigned the highest weight.

Table 2 Learned weights for each classifier in each cluster

Cluster #	1	2	3
Classifier 1	**0.4295**	−0.0475	**0.7301**
Classifier 2	0.2198	**0.6888**	−0.0019
Classifier 3	0.3536	0.3297	0.2811

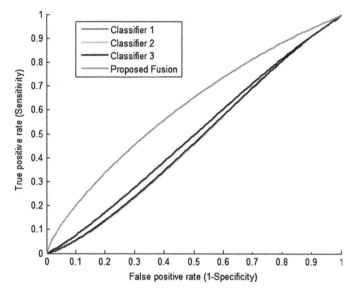

Fig. 1 Comparison of the three individual classifiers that use subsets of the features with the proposed fusion

Figure 1 displays Receiver Operating Characteristic (ROC) curve that compare the performance of the individual classifiers and the proposed fusion performances. As one can see, the proposed fusion outperforms the three individual classifiers.

To illustrate the local fusion ability of the proposed approach, we display the accuracy of the three individual classifiers (K-NN) for the 3 clusters. These accuracy values are obtained based on the crisp partition generated by the proposed algorithm, and testing the samples within each cluster independently. These results are displayed in Table 3 As it can be seen, the relative performances of the individual K-NN varies from one cluster to another. For instance in cluster 1, classifier 1 overcomes the classifier 2 and classifier 3. Consequently, for cluster 1 the most relevant classifier

Table 3 Per-cluster accuracy within each cluster obtained using the proposed Fusion

Cluster #	Cluster 1	Cluster 2	Cluster 3
Classifier 1	**0.9066**	0.7923	**0.9637**
Classifier 2	0.8934	**0.8239**	0.9592
Classifier 3	0.8574	0.7878	0.9589

is classifier 1. Thus, the highest aggregation weight is assigned to this classifier as shown in Table 2. Similarly, in cluster 2, the highest weights is assigned to classifier 2.

4 Conclusion

In this paper we have proposed a novel approach of automatic insult detection in social network comments. This approach relies on a local multi-classifier fusion method. Specifically, it categorizes the feature space into homogeneous clusters where a linear aggregation of the different classifier outputs yields a more accurate decision. The clustering process generates a possibilistic membership degree that represents the typicality, and is used to identify and discard noise frames. Moreover, the proposed algorithm provides optimal fusion parameters for each context. The initial experiments have shown that the fusion approach outperforms the individual classifiers performance. In order to overcome the need to specify the number of clusters apriori, a nice property of the possibilistic approach to generate duplicated clusters can be investigated in order to find the optimal number of clusters in an unsupervised manner.

Acknowledgments This work was supported by the Research Center of College of Computer and Information Sciences, King Saud University (Project RC131013). The authors are grateful for this support.

References

1. Spertus, E., Smokey: Automatic recognition of hostile messages. In: Proceedings of the Ninth Conference on Innovative Applications of Artificial Intelligence, pp. 1058–1065 (1997)
2. Mahmud, A., Ahmed, K.Z., Khan, M, Detecting flames and insults in text. In: Proceedings of the Sixth International Conference on Natural Language Processing (2008)
3. Razavi, A.H., Inkpen, D., Uritsky, S., Matwin, S., Offensive language detection using multi-level classification. In: Proceedings of the 23rd Canadian Conference on Artificial Intelligence, pp. 16–27 (2010)
4. Xiang, G., Hong, J., & Rosé, C. P. , Detecting Offensive Tweets via Topical Feature Discovery over a Large Scale Twitter Corpus, Proceedings of The 21st ACM Conference on Information and Knowledge Management, Sheraton, Maui Hawaii, October 29–November 2, (2012)
5. Xiang, G., Fan, B., Wang, L., Jason I., Carolyn, H., Rose, P., Detecting Offensive Tweets via Topical Feature Discovery over a Large Scale Twitter Corpus, Proceeding of the 21st ACM international conference on Information and knowledge management (CIKM '12), pp. 1980–1984 (2012)
6. Namburu, S.M., Tu,H., Luo, J., Pattipati, K.R., Experiments on Supervised Learning Algorithms for Text Categorization. International Conference, IEEE Computer Society, pp.1–8 (2005)
7. Liu, B., Zhang, L.: A survey of opinion mining and sentiment analysis. In: Aggarwal, C., Zhai, C. (eds.) Mining Text Data. Springer, Berlin (2011)
8. Lewis, D., Knowles, K.: Threading electronic mail: a preliminary study. Inf. Process. Manag. 33(2), 209–217 (1997)

9. Cohen, W., Learning rules that classify e-mail. AAAI Conference (1996)
10. de Carvalho, V.R., Cohen, W., On the collective classification of email "speech acts", ACM SIGIR Conference (2005)
11. Sahami, M., Dumais, S., Heckerman, D., Horvitz, E., A Bayesian approach to filtering junk e-mail. AAAI Workshop on Learning for Text Categorization. Technical Representation WS-98-05, AAAI Press. http://robotics.stanford.edu/users/sahami/papers.html
12. Bi, Y., Bell, D., Wang, H., Guo, G., Guan, J.: Combining multiple classifiers using dempster's rule for text categorization. Appl. Artif. Intell. **21**(3), 211–239 (2007)
13. Kuncheva, L.I.: Combining Pattern Classifiers. Wiley, New York (2004)
14. Sirlantzis, K., Hoque,S., Fairhurst, M. C., Trainable multiple classifier schemes for handwritten character recognition. In: Proceedings of the 3rd International Workshop on Multiple Classifier Systems, pp. 319–322, Cagliari, Italy (2002)
15. Huenupan, F., Yoma, N.B., Molina, C., Garreton, C.: Confidence based multiple classifier fusion in speaker verification. Pattern Recognit. Lett. **29**(7), 957–966 (2008)
16. Minsky, M.: Logical versus analogical or symbolic versus connectionist or neat versus scruffy. AI Mag. **12**(2), 34–51 (1991)
17. Hansen, L.K., Salamon, P.: Neural network ensembles. IEEE Trans. Pattern Anal. Mach. Intell. **12**(10), 993–1001 (1990)
18. Woods, K., Kegelmeyer Jr, W.P., Bowyer, K.: Combination of multiple classifiers using local accuracy estimates. IEEE Trans. Pattern Anal. Mach. Intell. **19**(4), 405–410 (1997)
19. Kuncheva, L., Clustering-and-selection model for classifier combination. In: Proceedings of Fourth International Conference on Knowledge-Based Intelligent Engineering Systems and Allied Technologies, vol. 1, pp. 185–188 (2000)
20. Liu, R., Yuan, B.: Multiple classifiers combination by clustering and selection. Information Fusion, pp. 163–168. Elseiver, New York (2001)
21. Frigui, H., Zhang, L., Gader, P.D., Ho, D., Context-dependent fusion for landmine detection with ground penetrating radar. In: Proceedings of the SPIE Conference on Detection and Remediation Technologies for Mines and Minelike Targets, Orlando, FL, USA, 2007
22. Abdallah, A.C.B., Frigui, H., Gader, P.D.: Adaptive Local Fusion With Fuzzy Integrals. IEEE T. Fuzzy Syst. **20**(5), 849–864 (2012)
23. Bezdek, J.C.: Pattern Recognition with Fuzzy Objective Function Algorithms. Plenum Press, New York (1981)
24. Krishnapuram, R., Keller, J.: A possihilistic approach to clustering. IEEE Trans. Fuzzy Syst. **1**, 98–110 (1993)
25. http://www.kaggle.com/c/detecting-insults-in-social-commentary/prospector#169 (2013)
26. Salton, G., Buckley, C.: Term-weighting approaches in automatic text retrieval. Inform. Process. Man. **24**(5), 513–523 (1988). Also reprinted in Sparck Jones and Willett [1997], pp. 323–328

What Information in Software Historical Repositories Do We Need to Support Software Maintenance Tasks? An Approach Based on Topic Model

Xiaobing Sun, Bin Li, Yun Li and Ying Chen

Abstract Mining software historical repositories (SHR) has emerged as a research direction Sun, over the past decade, which achieved substantial success in both research and practice to support various software maintenance tasks. Use of different types of SHR, or even different versions of the software project may derive different results for the same technique or approach of a maintenance task. Inclusion of unrelated information in SHR-based technique may lead to decreased effectiveness or even wrong results. To the best of our knowledge, few focus is on this respect in the SE community. This paper attempts to bridge this gap and proposes a preprocess to facilitate selection of related SHR to support various software maintenance tasks. The preprocess uses the topic model to extract the related information from SHR to help support software maintenance, thus improving the effectiveness of traditional SHR-based technique. Empirical results show the effectiveness of our approach.

Keywords Software historical repositories · Topic model · Information retrieval · Software maintenance

1 Introduction

Software maintenance has been recognized as the most difficult, costly and labor-intensive activity in the software development life cycle [21]. Effectively supporting software maintenance is essential to provide a reliable and high-quality evolution

X. Sun (✉) · B. Li · Y. Li · Y. Chen
School of Information Engineering, Yangzhou University, Yangzhou, China
e-mail: xbsun@yzu.edu.cn

B. Li
e-mail: lb@yzu.edu.cn

Y. Li
e-mail: liyun@yzu.edu.cn

Y. Chen
e-mail: chenying@yzu.edu.cn

© Springer International Publishing Switzerland 2015
R. Lee (ed.), *Computer and Information Science*, Studies in Computational Intelligence 566, DOI 10.1007/978-3-319-10509-3_3

of software systems. However, the complexity of source code tends to increase as it evolves. Recently, the field of software engineering (SE) focused on this field by mining the repositories related to a software project, for example, source code changes, bug repository, communication archives, deployment logs, execution logs [12]. Among these information, software historical repositories (*SHR*) such as source control repositories, bug repositories, and archived communications record information about the evolution and progress of a project. The information in *SHR* have been analyzed to support various software maintenance tasks such as impact analysis, bug prediction, software measures and metrics, and other fields [2, 12, 16, 23, 27]. All these studies have shown that interesting and practical results can be obtained from these historical repositories, thus allowing maintainers or managers to better support software evolution and ultimately increase its quality.

The mining software repositories field analyzes and explores the rich data available in *SHR* to uncover interesting and actionable information about software systems and projects [12]. The generated information can then be used to support various software maintenance tasks. These software maintenance tasks are traditional *SHR*-based techniques, which directly used the information in *SHR* in support of these software maintenance tasks without any filtering process. In practice, use of different types of software repositories, or even different versions of the software project may derive different results for the same technique or approach of a maintenance task [12, 13]. Sometimes, appropriate selection of the information in software repositories may obtain expected effectiveness for some techniques. However, inclusion of unrelated information in *SHR* for practical analysis may lead to decreased effectiveness. Hence, the main research question comes out:

"What information in *SHR* should be included to support software maintenance tasks?"

To the best of our knowledge, there is still few work to address this issue. In this paper, we focus on this respect, and attempt to facilitate selection of related *SHR* to support various software maintenance tasks. In the software repositories, the data can be viewed as unstructured text. And topic model is one of the popular ways to analyze unstructured text in other domains such as social sciences and computer vision [4, 15], which aims to uncover relationships between words and documents. Here, we proposed a preprocess before directly using *SHR*, which uses the topic model to help select the related information from *SHR*. After the preprocess, the effectiveness of traditional *SHR*-based techniques for software maintenance tasks is expected to be improved.

The rest of the paper is organized as follows: in the next section, we introduce the background of *SHR* and the topic model. Section 3 presents our approach to select the necessary information from *SHR*. In Sect. 4, empirical evaluation is conducted to show the effectiveness of our approach. Finally, we conclude and show some future work in Sect. 5.

2 Background

As our approach is to use topic model to facilitate selection of appropriate information from *SHR* to support software maintenance tasks. In this section, we introduce some background about *SHR* and the topic model.

2.1 Software Historical Repositories

Mining software repositories (MSR) has emerged as a research direction over the past decade, which has achieved great success in both research and practice [12]. Software repositories contain a wealth of valuable information about software projects such as historical repositories, runtime repositories, and code repositories [12]. Among these repositories, software historical repositories (*SHR*) record information about the evolution and progress of a project. *SHR* collect important historical dependencies between various software artifacts, such as functions, documentation files, and configuration files. When performing software maintenance tasks, software practitioners can depend less on their intuition and experience, and depend more on historical data. For example, developers can use this information to propagate changes to related artifacts, instead of using only static or dynamic program dependencies, which may fail to capture important evolutionary and process dependencies [11].

SHR mainly include source control repositories, bug repositories, communication archives. A description of these repositories is shown in Table 1. The amount of these information will become lager and larger as software evolves. These information

Table 1 Software historical repositories

Software historical repositories	Description
Source control repositories	These repositories record the information of the development history of a project. They track all the changes to the source code along with the meta-data associated with each change, for example, who and when performed the change and a short message describing the change. CVS and subversion belong to these repositories.
Bug repositories	These repositories track the resolution history of bug reports or feature requests that are reported by users and developers of the projects. Bugzilla is an example of this type of repositories.
Communication archives	These repositories track discussions and communications about various aspects of the project over its lifetime. Mailing lists and emails belong to the communication archives.

can be used to support various software maintenance tasks, such as change impact analysis, traceability recovery [12, 14]. These traditional *SHR*-based techniques directly used the information in the *SHR* to perform software maintenance. However, with the evolution of software, some information may be outdated. Therefore, not all historical information are useful to support software maintenance. In this paper, we focus on selection of useful or related information from *SHR* to support practical software maintenance tasks.

2.2 Topic Model

As information stored in *SHR* is mostly unstructured text, researchers have proposed various ways to process such unstructured information. An increasingly popular way is to use topic models, which focus on uncovering relationships between words and documents [1]. Topic models were originated from the field of natural language processing and information retrieval to index, search, and cluster a large amount of unstructured and unlabeled documents [25]. A topic is a collection of terms that co-occur frequently in the documents of the corpus. The most used topic models in software engineering community are Latent Semantic Indexing (LSI) and Latent Dirichlet Allocation (LDA) [25]. These two topic models have been applied to support various software engineering tasks: feature location, change impact analysis, bug localization , and many others [17, 19, 25]. The topic models require no training data, and can well scale to thousands or millions of documents. Moreover, they are completely automated.

Among the two topic models, LDA is becoming increasing popular because it models each document as a mixture of K corpus-wide topics, and each topic as a mixture of the terms in the corpus [5]. More specifically, it means that there are a set of topics to describe the entire corpus, each document can contain more than one of these topics, and each term in the entire repository can be contained in more than one of these topic. Hence, *LDA* is able to discover a set of ideas or themes that well describe the entire corpus. *LDA* is a probabilistic statistical model that estimates distributions of latent topics from textual documents [5]. It assumes that the documents have been generated using the probability distribution of the topics, and that the words in the documents were generated probabilistically in a similar way [5]. With *LDA*, some latent topics can be mined, allowing us to cluster them on the basis of their shared topics. In this paper, we use *LDA* to extract the latent topics from various software artifacts in *SHR*.

3 Approach

Our main focus in this paper is to provide an effective way to automatically extract related information from the *SHR* to provide support of software maintenance tasks. The process of our approach is shown in Fig. 1, which can be seen as a preprocess to traditional *SHR*-based techniques. Given a maintenance task or request, we need some related information in *SHR* to well support comprehension, analysis and implementation of this request. The data source of our approach includes a maintenance request, *SHR* and current software. We extract current software from *SHR* here because the current software usually needs some necessary changes to accomplish this change request. As the data source can be seen as unstructured text, we use *LDA* on these data source to extract the latent topics in them. Then, we compare the similarity among the topics extracted from different data source, and ultimately produce the related software repositories which are related to the maintenance request and current software. For example, in Fig. 1, when there exists a bug repository which has similar topics with the software maintenance request, we can consider this bug repository as related data source to analyze the current software maintenance request. In addition, there is also a *feedback* from this bug repository to its corresponding version repository in source control repositories. More specifically, the corresponding version evolution corresponding to this bug repository is also considered to be a useful data source for analyzing the request in the current software. Hence, the extracted related information from *SHR* includes the related bug repository, useful communication archive, and some evolutionary code versions related to this maintenance request and the current software.

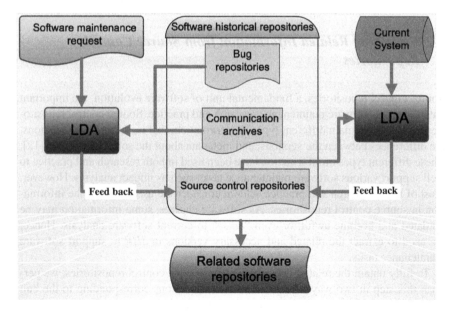

Fig. 1 Process of our approach

3.1 Extracting Related Information from Bug Repositories and Communication Archives

First, we extract the related information from bug repositories and communication archives to see what information is related to the maintenance request. We use the *LDA* model to extract the latent topics in these data source, i.e., maintenance request, bug repositories, and communication archives. The topics extracted from these data source can be represented with a vector (for maintenance request) and matrices (for bug repositories and communication archives). Then, we compute the similarity between the vector and these two matrices, respectively. There are many approaches to compute the similarity result, for example, distance measures, correlation coefficients, and association coefficients [22]. Any similarity measure can be used in our approach. According to the similarity results, the bug repositories and communication archives which are similar to the maintenance request can be extracted as useful data source. These extracted information can be helpful to guide the change analysis. For example, when the maintenance request is to fix a bug and if there is an existing bug report similar to this bug fixing request, we can use the information in related bug repositories and communication archives to see "who fixes the bug?", "how to fix the bug?", etc.

3.2 Extracting Related Information from Source Control Repositories

Source control repositories, a fundamental unit of software evolution, are important data source in software maintenance research and practice. Source control repositories include three main different types of information, that is, the software versions, the differences between the versions, and metadata about the software change [12]. These different types of information have been used in both research and practice to well support various software maintenance tasks such as impact analysis. However, most of these research and practice seldom consider the usefulness of the information in source control repositories. As software evolves, some information may be outdated and become awful, or even "*noise*" to current software analysis. Hence, we need to extract the related and necessary versions in *SHR* to support software maintenance tasks.

To fully obtain the related information from source control repositories, we perform this step in two ways. First, we select the versions corresponding to the bug repositories and communication archives from the previous step, where we have obtained the bug repositories and communication archives related to the maintenance request. In addition, we can also obtain the **consecutive** source code versions corresponding to the bug repositories and communication archives.[1] Thus, this kind of source code versions is produced based on the maintenance request. Second, we extract the source versions similar to current software. We extract the latent topics from current software and represent them as a vector. Then, we compute the similarity result between the current software vector and the matrices (for source control repositories). Here, we also extract consecutive versions which are related to the current software.

Until here, all the related information have been extracted from various *SHR*. We believe that such a preprocess on [12] can effectively improve the effectiveness of traditional software maintenance activities which are not preprocessed or filtered.

4 Evaluation

Our approach aims to improve traditional *SHR*-based software maintenance techniques after filtering the unrelated information without decreasing the completeness of the results.

4.1 Empirical Setup

We conduct our evaluation on an existing benchmark built by Poshyvanyk et al.[2] We select three Java subject programs from open projects for our studies. The first subject program is *ArgoUML*, which is an open source UML modeling tool that supports standard UML 1.4 diagrams. The second subject is *JabRef*, which is a graphical application for managing bibliographical databases. The final subject is *jEdit*, which is a text editor written in Java. For each subject program, we used four concecutive versions ($V0 \rightarrow V3$) for study. Then, we compare the effectiveness of our approach with traditional software maintenance techniques on this benchmark. Here, we use change impact analysis as the representation of one of the software maintenance techniques for study [14]. Change impact analysis is a technique used to identify the potential effects caused by software changes [6, 14]. In our study, we employ *ROSE* (Reengineering of Software Evolution) tool to represent the *SHR*-based CIA [28]. *ROSE* is shown to be effective for change impact prediction [24], and it applies data mining to version histories to guide impact analysis. The input of *ROSE* can be coarser file or class level changes, or finer method-level changes. Its output is the corresponding likely impacted entities at the same granularity level as the change set. In this paper, the chosen granularity level is class level.

---

[1] Here we need consecutive versions since many software maintenance tasks are performed based on the differences between consecutive versions in source control repositories.

[2] http://www.cs.wm.edu/semeru/data/msr13/.

To show the effectiveness of CIA, we used *precision* and *recall*, two widely used metrics of information retrieval [26], to validate the *accuracy* of the CIA techniques. They are defined as follows:

$$P = \frac{|Actual \quad Set \quad \cap \quad Estimated \quad Set|}{|Estimated \quad Set|} \times 100\%$$

$$R = \frac{|Actual \quad Set \quad \cap \quad Estimated \quad Set|}{|Actual \quad Set|} \times 100\%$$

Actual Set is the set of classes which are really changed to fix the bugs for the system. *Estimated Set* is the set of classes potentially impacted by the *change set* based on *ROSE*. The change set is composed of a set of classes used to fix each bug. The change set is mined from their software repositories. Specific details on the process of the identification of the bug reports and changed classes can refer to [20]. With the change set, we applied *ROSE* to compute the *Estimated Set*. Then, *precision* and *recall* values are computed based on the *Actual Set* and *Estimated Set*. Here, *ROSE* is performed twice, one is on the original software historical repositories, i.e., *ROSE*; the other is on the extracted repositories preprocessed by our approach, i.e., *ROSE'*.

4.2 Empirical Results

We first see the precision results of *ROSE* before (*ROSE*) and after (*ROSE'*) software historical data filtering. The results are shown in Fig. 2. From the results, we see that all the precision values of *ROSE* are improved after software historical data filtering. It shows that there is indeed some unrelated information in software historical repositories for software maintenance and evolution. Hence, we should filter these unrelated software historical information to improve the precision of change impact analysis.

In addition, during the process of filtering the software historical data, there may be some related information which is filtered by our approach, so we need to see

Fig. 2 Precision of ROSE before and after information filtering

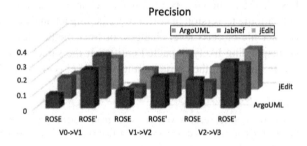

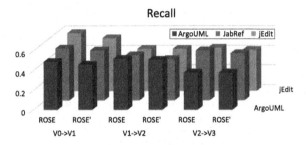

Fig. 3 Recall of ROSE before and after information filtering

whether the recall of *ROSE* is seriously decreased after information filtering. The recall results are shown in Fig. 3. It shows that most of the recall values are decreased after information filtering process. However, the degree of the deceasing values is not big. Hence, we can obtain that only a small amount of related information is filtered during the process, which has few effect on software maintenance and evolution.

5 Conclusion and Future Work

This paper proposed a novel approach which can improve the effectiveness of traditional *SHR*-based software maintenance tasks. Our approach extracted related information from *SHR* using the topic model, i.e., LDA. The generated software repositories can eliminate the information that are outdated during software evolution, thus improving the effectiveness of the traditional *SHR*-based software maintenance tasks. Finally, this paper evaluated the proposed technique and showed its effectiveness.

Though we have shown the effectiveness of our approach through real case studies based on *ROSE*, it can not indicate its generality for other real environment. And we will conduct experiments on more real programs to evaluate the generality of our approach. In addition, we would like to study whether the effectiveness of other software maintenance tasks (e.g., feature location [8, 9], bad smell detection [10, 18], traceability recovery [3, 7], etc.) can be improved based on the *SHR* preprocessed by our approach.

Acknowledgments The authors would like to thank anonymous reviewers who make the paper more understandable and stronger. This work is supported partially by the Natural Science Foundation of the Jiangsu Higher Education Institutions of China under Grant No. 13KJB520027, partially by National Natural Science Foundation of China under Grant No. 61402396 and No. 61472344, partially by the Innovative Fund for Industry-Academia-Research Cooperation of Jiangsu Province under Grant No. BY2013063-10, and partially by the Cultivating Fund for Science and Technology Innovation of Yangzhou University under Grant No. 2013CXJ025.

References

1. Anthes, G.: Topic models versus unstructured data. Commun. ACM **53**(12), 16–18 (2010)
2. Antoniol, G., Huffman Hayes, J., Gaël Guéhéneuc, Y., Di Penta, M.: Reuse or rewrite: Combining textual, static, and dynamic analyses to assess the cost of keeping a system up-to-date. In: 24th IEEE International Conference on Software Maintenance, pp. 147–156 (2008)
3. Antoniol, G., Canfora, G., Casazza, G., De Lucia, A., Merlo, E.: Recovering traceability links between code and documentation. IEEE Trans. Software Eng. **28**(10), 970–983 (2002)
4. Barnard, K., Duygulu, P., Forsyth, D.A., de Freitas, N., Blei, D.M., Jordan, M.I.: Matching words and pictures. J. Mach. Learn. Res. **3**, 1107–1135 (2003)
5. Blei, D.M., Ng, A.Y., Jordan, M.I.: Latent dirichlet allocation. J. Mach. Learn. Res. **3**, 993–1022 (2003)
6. Bohner, S., Arnold, R.: Software Change Impact Analysis. IEEE Computer Society Press, Los Alamitos (1996)
7. De Lucia, A., Di Penta, M., Oliveto, R., Panichella, A., Panichella, S.: Applying a smoothing filter to improve ir-based traceability recovery processes: An empirical investigation. Inf. Softw. Technol. **55**(4), 741–754 (2013)
8. Dit, B., Revelle, M., Gethers, M., Poshyvanyk, D.: Feature location in source code: a taxonomy and survey. J. Softw. Evol. Process **25**(1), 53–95 (2013)
9. Dit, B., Revelle, M., Poshyvanyk, D.: Integrating information retrieval, execution and link analysis algorithms to improve feature location in software. Empir. Softw. Eng. **18**(2), 277–309 (2013)
10. Fontana, F.A., Braione, P., Zanoni, M.: Automatic detection of bad smells in code: an experimental assessment. J. Object Technol. **11**(2), 5:1–5:38 (2012)
11. Hassan, A.E., Holt, R.C.: Predicting change propagation in software systems. In: 20th International Conference on Software Maintenance, pp. 284–293 (2004)
12. Kagdi, H.H., Collard, M.L., Maletic, J.I.: A survey and taxonomy of approaches for mining software repositories in the context of software evolution. J. Softw. Maintenance **19**(2), 77–131 (2007)
13. Kagdi, H.H., Gethers, M., Poshyvanyk, D.: Integrating conceptual and logical couplings for change impact analysis in software. Empir. Softw. Eng. **18**(5), 933–969 (2013)
14. Li, B., Sun, X., Leung, H., Zhang, S.: A survey of code-based change impact analysis techniques. Softw. Test., Verif. Reliab. **23**(8), 613–646 (2013)
15. Li, D., Ding, Y., Shuai, X., Bollen, J., Tang, J., Chen, S., Zhu, J., Rocha, G.: Adding community and dynamic to topic models. J. Informetrics **6**(2), 237–253 (2012)
16. Mockus, A., Fielding, R.T., Herbsleb, J.D.: Two case studies of open source software development: Apache and mozilla. ACM Trans. Softw. Eng. Methodol. **11**(3), 309–346 (2002)
17. Nguyen, A.T., Nguyen, T.T., Al-Kofahi, J., Nguyen, H.V., Nguyen, T.N.: A topic-based approach for narrowing the search space of buggy files from a bug report. In: Proceedings of the IEEE/ACM International Conference on Automated Software Engineering, pp. 263–272 (2011)
18. Palomba, F., Bavota, G., Di Penta, M., Oliveto, R., De Lucia, A., Poshyvanyk, D.: Detecting bad smells in source code using change history information. In: IEEE/ACM International Conference on Automated Software Engineering, pp. 268–278 (2013)
19. Panichella, A., Dit, B., Oliveto, R., Di Penta, M., Poshyvanyk, D., De Lucia, A.: How to effectively use topic models for software engineering tasks? an approach based on genetic algorithms. In: 35th International Conference on Software Engineering, pp. 522–531 (2013)
20. Poshyvanyk, D., Marcus, A., Ferenc, R., Gyimothy, T.: Using information retrieval based coupling measures for impact analysis. Empir. Softw. Eng. **14**(1), 5–32 (2009)
21. Schneidewind, N.F.: The state of software maintenance. IEEE Trans. Softw. Eng. **13**(3), 303–310 (1987)
22. Shtern, M., Tzerpos, V.: Clustering methodologies for software engineering. Adv. Softw. Eng. **2012**, 18. doi:10.1155/2012/792024 (2012)

23. Sliwerski, J., Zimmermann, T., Zeller, A.: When do changes induce fixes? ACM SIGSOFT Softw. Eng. Notes **30**(4), 1–5 (2005)
24. Sun, X., Li, B., Li, B., Wen, W.: A comparative study of static cia techniques. In: Proceedings of the Fourth Asia-Pacific Symposium on Internetware, pp. 23 (2012)
25. Thomas, S.W.: Mining software repositories using topic models. In: Proceedings of the 33rd International Conference on Software Engineering, pp. 1138–1139 (2011)
26. van Rijsbergen, C.J.: Information Retrieval. Butterworths, London (1979)
27. Zimmermann, T., Weißgerber, P., Diehl, S., Zeller, A.: Mining version histories to guide software changes. IEEE Trans. Softw. Eng. **31**(6), 429–445 (2005)
28. Zimmermann, T., Zeller, A., Weissgerber, P., Diehl, S.: Mining version histories to guide software changes. IEEE Trans. Softw. Eng. **31**(6), 429–445 (2005)

Evaluation Framework for the Dependability of Ubiquitous Learning Environment

Manel BenSassi, Mona Laroussi and Henda BenGhezela

Abstract Ubiquitous learning environment confronts a set of challenges. First, how to make use of technology without losing pedagogical aspect of learning. Second, how to specify evaluation dimensions, aspects and criteria—what to evaluate, how we evaluate and what to take into account to evaluate. In order to response to these questions, we begin by reviewing the literature to determine challenges and evaluation's dimensions introduced by the ubiquitous learning environment. And then we introduce a proposed framework of evaluation ubiquitous learning that treat the issue of how considering contextual dimension from a technological point of view. This framework is considered in our researches that are interested in developing ubiquitous learning environments based on wireless and sensor technologies. Finally, we detail how we exploit this framework to evaluate a realistic case study.

Keywords Learning scenario · Pre evaluation · Dependability of ubiquitous learning environment · Formal modeling · Contextual evaluation

1 Introduction

As mobile and embedded computing devices become more ubiquitous, it is becoming obvious that the interactions between users and computers must evolve. Learning environments and applications need to become increasingly not only autonomous and invisible but also, dependable and reliable.

Dependability is an important pre-requisite for ubiquitous learning [1] and should be evaluated sufficiently and at early stage to respond user's requirement. In fact, the

M. BenSassi (✉) · M. Laroussi · H. BenGhezela
RIADI GDL, Ensi, University of Manouba, Manouba, Tunisia
e-mail: bensassi.manel@gnet.tn

M. Laroussi
e-mail: mona.laroussi@planet.tn

H. BenGhezela
e-mail: hhbg.hhbg@gnet.tn

© Springer International Publishing Switzerland 2015
R. Lee (ed.), *Computer and Information Science*, Studies in Computational
Intelligence 566, DOI 10.1007/978-3-319-10509-3_4

learning environment had to operate as learners and designers expect and that it will not fail in normal [2].

Many researchers have defined the dependability in slightly different way. Dependability is an orthogonal issue that depends on QoS. We consider its original meaning as defined in [3, 4]: *Dependability is the quality of the delivered such that reliance can justifiably be placed on this service*. It could be typically divided into a number of aspects namely Avizienis et al. [4]: (functional) correctness, safety, security, reliability, availability, transparency and traceability. Each dimension includes analysis techniques, assessment methods and measures. Ubiquitous learning environments come with requirements from all aspects of dependability. The reason for this lies in the nature of these ubiquitous scenarios itself. Typically these environments are very tightly connected with specific users.

The dependability's evaluation is an important activity as well as the evaluation of the executed learning scenario. Post-evaluation techniques and approaches take much time and cost, and an early preventive evaluation could be one of the key factors to cost effective ubiquitous learning development.

The basic question is: *How to evaluate the learning functionalities in their context of use at early stage in order to ensure the dependability of the conceived scenario?* To answer to this question, we first summarize our definitions. We define learning scenario as technological environment consisting of one or more activities, correlated together offering a complete scenario of information and communication services required for supporting learning. The ubiquitous learning can be defined as *environment that provides an interoperable, mobile, and seamless learning architecture to connect, integrate, and share three major dimensions of learning resources: learning collaborators, learning contents, and learning services. Ubiquitous learning is characterized by providing intuitive ways for identifying right learning collaborators, right learning contents and right learning services in the right place at the right time* [5].

This paper proposes an early preventive evaluation framework: ReStart-Me (**Re**-engineering educational **S**cenario based on **t**imed **a**utomata and **tr**acks **t**reatment for **M**alleable learning **e**nvironment) that intends to reduce the time and cost through using test cases as a means of the evaluation. Test cases are developed in the process of the leaning scenario design and used to test the target scenario. ReStart-Me checks formally inclusion relation between dependability requirements and test cases. If the formal checking succeeds, then we can assure that the dependability requirements are well implemented in the ubiquitous learning environment.

This paper is organised as follows. The following section will be devoted to define challenges of ubiquitous learning scenario. Section 3 presents different dimensions that should be evaluated in order to satisfy dependability's requirement. Section 4 focus on our evaluation methodology based on the timed automata modeling and formal verification properties while an experiment of our work through ubiquitous learning scenario is included in Sect. 5. Section 6 concludes the paper and suggests future research directions.

2 Ubiquitous Learning Scenario Challenges

According to Yahya [6]: *"we move from conventional learning to electronic-learning (e-learning) and from e-learning to mobile-learning (m-learning) and now we are shifting to u-learning"*.

In this section, we present a review of existent learning scenarios paradigms through a comparative study based on multicriteria classification. This review aims to discuss issues on ubiquitous learning.

In order to analyse and effectively understand current Learning paradigm, a classification still remains essential. In fact, we propose four criteria from the combination of researchers' ideas [6, 7]:

- **Level of embeddedness**: This criterion reflects the degree of technology's visibility to the user when he interacts with the external environment.
- **Level of mobility**: The learner is mobile and uses mobile technolgies to learn.
- **Level of interactivity**: The learner can interact with peers, teachers and experts efficiently and effectively through different media.
- **Level of context-awareness**: The environment can adapt to the learners real situation to provide adequate information.

Table 1 summarizes our comparative study based on criteria described above. As we show in Table 1, the responsibility of the ubiquitous learning infrastructure with respect to applications, according to [8], includes supporting application requirements such as context awareness, adaptation, mobility, distribution and interoperability; facilitating the rapid development and deployment of software components; providing component discovery services; and providing scalability.

Ubiquitous learning scenario is a context aware environment that is able to sense the situation of learners and provide various adaptive functionalities and supports. Many researchers have been investigating the development of such new learning environments. Nevertheless, the higher cost of its implementation and the complexity of their context emphasize the need to evaluate their correctness and to detect functional deficiencies at early stage before implementation phase.

Table 1 A comparative study of learning paradigm

Learning paradigm	Level of embeddedness	Level of mobility	Level of interactivity	Level of context-awareness
E-learning	+	+	+	+
Mobile learning	+	+++	++	++
Pervasive learning	+++	++	+++	+++
Ubiquitous learning	+++	+++	+++	+++

legend: +: Low ++: Medium +++: High

3 What Can We Evaluate in Ubiquitous Learning Scenario?

As computers have become more rapid and powerful, educational software has flourished and there are numerous claims conceived by designers. Thus evaluation software is important so that teachers can make an appropriate choice of learning scenarios and which are suitable to the teaching and learning context.

According to [9], we can evaluate these following dimensions in a learning scenario: content, technical support, learning process (see Fig. 1). Each dimension includes a number of sub categories and criteria that could be considered in the evaluation process. All these aspects are equally important, as the learning activity has to be simultaneously pedagogically and technically sound.

Nevertheless, ubiquitous learning infrastructure is centered on a high-level conceptual model consisting of devices, users, software components and user interfaces. Then, we propose that the learning activity's context should also be considered for evaluation. The context is a set of evolutive elements appropriate to the interaction between learner and learning application including the learner and the learning environment themselves. Figure 1 presents in a diagram the different aspects that we can combine and include in the evaluation framework.

Ubiquitous learning could be evaluated from different viewpoints: The first one is a technical oriented perspective. The content suitable for m-learning needs to be

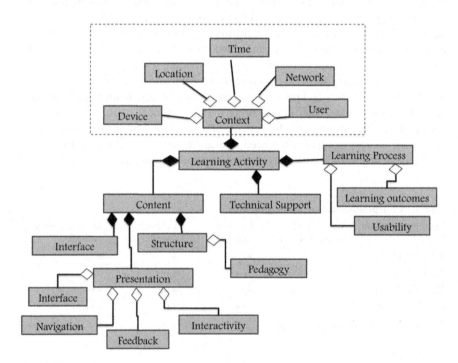

Fig. 1 Different features that could be evaluate in the learning scenario

available and adequate to the learning environment. The second one is pedagogical oriented perspective. It points out that m-learning develops new skills and approaches to ensure the pedagogical effectiveness. We consider that the context of ubiquitous environment should be also considered in evaluation process. The features of activity's context are: location, network, user, time and device.

- **User dimension**: A learning scenario may imply different actors or entities found in the given environment. Thus according to the tasks that learners are asked to accomplish, they may need to interact with one or many actors, each playing various roles in the learning process. The malleable learning environment may include these types of actors or entities:

1. *Instructors* (*I*): This role is assigned to a teacher, a professor, a laboratory assistant or anyone involved in the management of the learning scenario.
2. *Learner* (*L*): Learners are the central entities of the learning scenario as the whole learning process is hinged upon them.
3. *Group* (*G*): A group is a set of learners members can cooperate and communicate between them.
4. *Smart entities* (*E*): Like capture, smart sensors in pervasive environment.

These different entities make it possible to illustrate the interactions within the learning management system. Thus our conceptual framework is designed around these entities.

- **The device dimension**: We consider that it is important capabilities of user's device, especially hardware attributes, for ubiquitous learning due to the fact they have a big impact on learning scenario execution.
- **The network and connectivity dimension**: Nowadays mobile device might be connected to the "Net" via many technologies: GPRS, UMTS, WiFi, 3G communication, etc. Mobile devices often have periods of disconnection that had to be considered in evaluation of the dependability of the learning scenario. The connectivity quality depends on user's location and mobility. According to the number of participants and the time the connection is established, a communication will be classified in one of many possible categories. In order to describe all possible categories, we will limit the scope of this paper to two types of consideration:

  - *Quantitative considerations*: Number of participants, the presence of a group or not, the type of established interaction:
    Unicast: This is the simplest and most basic type of communication. In this case, only two active entities are involved in the communication.
    Broadcast: This communication is used when the coordinator wants to reach all learners participating in the learning scenario.
    Multicast: Finally, multicast communication occurs when the instructor sends a message to a specific group of learners.
  - *Qualitative considerations*: Level of technology, the quality of signal in Wireless connection

- **Temporal dimension**: The designed learning scenario may have different temporal requirements that we classify according to the type of the used activities.

  - *Synchronous*: For this category, activities in learning scenario are usually running at the same time [10] and they must be mutually synchronized. Furthermore, different entities participating in the communication progress simultaneously.
  - *Asynchronous*: This type of time constraints is used when the entities participating in the activity are not present simultaneously at the same time [11].

4 Evaluation of Ubiquitous Learning Scenarios: Related Works and Issues

There have been several researches to develop methods to evaluate ubiquitous learning scenario. The most widely used methods are:

- Heuristic evaluation method based on principals and many categorised dimensions [12]. A wide range of methods has been developed to systematically evaluate the quality of information technologies.
- Evaluation methods based on Simulations [13] where some user problems are simulated in details, especially analysing each task from a cognitive point of view.
- Evaluation system based on tracks analysis requests both analytical and technical staff. The first group is responsible for defining scores for various features of the e learning scenario according to a specific set of evaluation coefficients. This team also specify the quality of the learning scenario. The technical staff develops the system or specify the evaluation framework to the mentioned scores [14, 15].
- Teaching test methods are based on an appropriate testing program that is suitable to different goals and characteristics of the teaching style. These methods implement the pre-test, the post-test of the learning scenario, and other steps to test the knowledge of students and their skills and finally determine the effectiveness of teaching scenario [16].

As shown, there are several evaluation studies. However, these methods vary widely in their evaluation scope, outcomes and techniques. In fact, in the one hand, there are generic methods that, although useful in theory, are not very applicable in practice, since they do not take into account the situatedness of the courseware evaluation, determined by the context of the learning scenario. In the other hand, some of these methods described above have a very specific target (cognitive overview). Nevertheless, with the growth in the use of mobile technologies and the learning scenario database, an increasing number of teachers want to reuse their scenario in different contexts.

Instead of adding to the already large number of checklists for learning scenario evaluation, we are attempting, in the following section, to address contextual evaluation based on formal modeling and verification of educational scenario. Next, the evaluation framework is discussed.

5 ReStart-Me: A Formal Evaluation Framework For U-Learning

5.1 ReStart-Me Challenges

In order to fill the gap between learning scenarios evaluation methods and ubiquitous learning scenarios, we present in this paper, a formal method to evaluate scenarios at an early stage: ReStart-Me (Re-engineering educational Scenario based on timed automata and tracks treatment for Malleable learning environment). This formal evaluation is based on automata theory and formal verification and aims to:

1. Avoid costly and time-consuming scenarios implementation or deployment
2. Simulate scenario execution on real time in order to check properties and to detect errors (such as deadlocks and liveness) and then to regulate scenario with timing constraints.

The evaluation framework is based on different kind of context: Actors, Location, Device, Network and connectivity. These elements are motivated by an empirical analysis of different definitions of learning context. In fact, these elements represent core features of any educational scenario supporting any learning approaches such as hybrid learning, mobile learning, ubiquitous learning, pervasive learning,...

Consequently, they represent interesting evaluation criteria in order to capture at early stage inconsistencies and design weakness.

5.2 Conceptual Framework

Figure 2 overviews the proposed scenario dependability evaluation framework for ubiquitous learning. ReStart-Me intends to reduce the cost of learning environment quality evaluation through using test cases. It uses formal checking techniques to ascertain inclusion relation between dependability requirements and test cases. Dependability requirements and test cases are modeled into timed automata. Then, they are transformed into temporal logical properties to formal checking methods.

The evaluation process contains essentially these three steps (see Fig. 2) inspired from [17]:

- **Step 0: Behavioural Modeling:** This phase provides informal or/and semiformal descriptions of both the learning environment and the correctness properties to be checked. Pedagogues and designers have to participate in this phase.
- **Step 1: Formal Modeling of Educational Scenario:** In this step, conceptual designers identify the entities to be considered. Designers have to successively refine it and decide which the final actors to be modelled are. The content of each process is created and redefined by modelling the dynamic behaviour. Then, we extend the obtained automaton with global and local clocks. We also define contextual constraints and correlation between different activities. By creating a

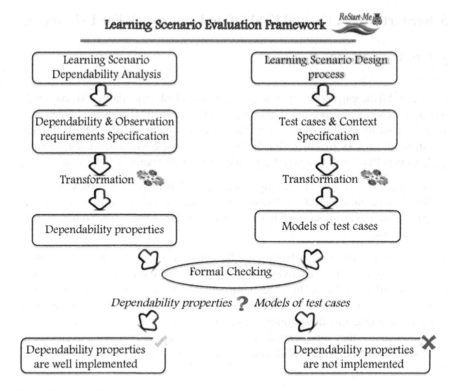

Fig. 2 Dependability evaluation framework for learning scenario

formal specification of the learning scenario, the designers are forced to make and
to define a detailed scenario analysis at early stage before its deployment into the
learning system.

- **Step 2: Tracks Simulation:** Through simulation, we can observe all possible
  interactions between automata corresponding to different entities involved in the
  learning scenario. This step generates simulated tracks that facilitate errors detection. A simulation is equivalent to an execution of the designed learning scenario.
  It gives some insight on how the model created behaves.
- **Step 3: Properties verification:** With formal verification, we check the correctness
  of the designed learning scenario. This process aims to build a remedial scenario
  and to help designer in the reengineering process by providing Errors and warning
  reports.

ReStart-Me can be applied with the help of different developing methods and
tools. To exemplify this, we have selected a tool that is well known, efficient and
provide us with most of what we need. The selected tool is a model checker that has
a modelling language, plus simulation and formal verification capabilities. We think
that our choice is well founded but is not essential. To illustrate the application of
ReStart-Me, we will use UPPAAL model checker [18].

6 Experiment and Results

6.1 Case Study

The objective of this study reported in this section is twofold: (i) to check practicality of the proposed framework when building an actual ubiquitous scenario (ii) and to evaluate qualitatively the benefits of such methodology. A team of researchers composed mainly by of five designers has prepared a textual description of the ubiquitous learning scenario.

6.1.1 Evaluation's Feature

Formal evaluation must pay attention first to dependability's aspect. Enhancing this idea, evaluation features can be defined in the form of questions concerning the evaluation methedology as shown in Table 2.

6.1.2 Learning Scenario Description

The conceptual designers' meetings have produced a scenario design document that describes the sequencing of different activities and rules of the ubiquitous learning scenario. To summarize, the learning scenario can be described as follows:

A high school decides to raise awareness of pupils on the effects of pollution on the environment by organizing a trial allowing the follow-up of the pupil's educational curriculum in the field of "Education about the environment". This trial enables pupils to learn through factual cases and to experiment various scenarios using mobile technologies.

The physical setting of this trial, where activities take place, is an ecological zone near by an industrial area. It is organized in the morning from 9:00 am to 13:00 am. In order to boost intra-group competition, students were divided in three groups under the supervision of their coach and each group consisted of six pupils.

Table 2 Evaluation features

Question	Response
What is being evaluated?	The ubiquitous learning scenario
What is the purpose of the evaluation?	The dependability of the contextual learning scenario
When should the evaluation be done?	At early stage (before learning scenario's deployment)
Who should perform the evaluation?	Conceptual designer
What is the type of evaluation?	A formal evaluation based on timed automata and model checking

Additionally, each group was divided in two subgroups of three students. The ultimate goal behind this clustering is to reinforce teamwork and collaboration within the individual subgroups and to make it a collaborative and challenging game that takes place in different locations.

The outdoor subgroup is equipped with a smart phone with a wireless connection. At the beginning of the first stage, a localization sensor localizes the out door's subgroup and a notification is sent to ask students to identify and take a photo of the QR-code stuck to a tree. Instantly, a text adapted to the pupils' level and pictures that visualize and describe the activities to accomplish in the current stage is displayed on the screen of the smart phone.

In the first stage, the outdoor subgroup can take a photo of a plant and search for a plant related groups and then share the photo and ask for help in identifying it.

After a pre-defined time, the subgroup will receive a stage-adapted quiz via automatic text message. Pupils need to write an answer using their smartphone and submit it. If the answer submitted by the group is not correct, the system sends an alert to the coach informing him/her that pupils need some support. The coach should send to them some hints.

In order to improve the coaching task, tutor decides that after three wrong attempts, the pupil is guided to start learning session by using his mobile device. The e-learning client allow the student to directly mash up widgets to create lesson structure and add powerful online test widgets, communication widget (chat, forum and personal messages), content scheduling widgets, communication tracking, announcements, content flows, cooperative content building widgets.

The student could drag from the widget repository and drops into the elearning client UI all the widgets needed for providing video, audio and other multimedia content. The session of learning take 30 min with GPRS connection and 40 min with EDGE connection (see Table 3).

Else, if the answer is correct the indoor subgroups will receive the list of activities of the second stage and will get joined by their corresponding outdoor subgroups that will hand over the picked plant samples. At the end, indoor and outdoor subgroups of each of the three groups should collaborate in drawing their own conclusions using the collaborative tool Google Docs and then present the outcome of their study about effects of pollution on environment.

Table 3 Context information (network connection quality)		Zone 1	Zone 2
	Type of network	EDGE	GPRS

6.2 Modelisation and Evaluation

6.2.1 Formal Modelisation

Figures 3 and 4 provide an overview of different automata modeled for the planned learning activities (outdoor activities, quiz, and lesson) and corresponding to student. We define a global variable "clock" named *Time* that gives idea about the duration of each activity. The timing constraints associated with locations are invariants. It gives a bound on how long these locations can be active. We also define other integer global variable *Power* to calculate the energy of battery of the smartphone.

In order to facilitate the learning scenario analysis, we model each activity separately. The idea is to define templates for activities that are instantiated to have a simulation of the whole scenario. The motivation for the use of templates is that the understanding, the share and the reuse of different components of the learning scenario become easier.

The whole scenario is modelled as a parallel composition of timed automata. An automaton may perform a transition separately or synchronise with another automaton (channel synchronisation) or it can be activated after a period of time through flags.

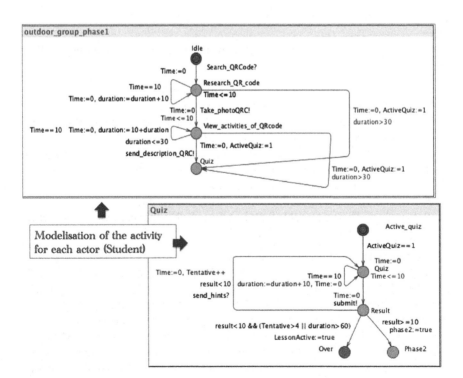

Fig. 3 The automata model of different activities for one actor (student)

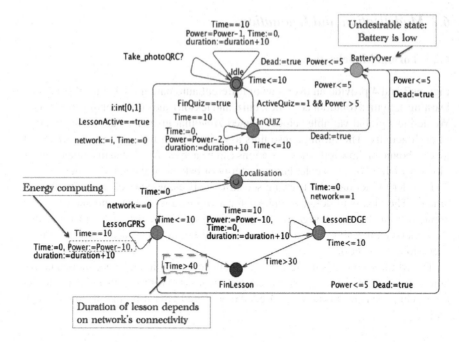

Fig. 4 The automaton model of the smartphone

Figure 4 shows a timed automaton modeling the behavior of the Smartphone. This device has several locations (or states):

1. *Idle*: This state is activated when the smartphone is switched on.
2. *InQuiz*: The learner is responding to the quiz's questions.
3. *Localisation*: This state is activated when the learner has to do his lesson. The smartphone had to be connected to the gived wireless network.
4. *LessonGPRS* and *LessonEDGE*: One of these two states is activated. The learner drags and drops into the elearning client UI all the widgets needed for providing video, audio and other multimedia content.
5. *FinLesson*: The student has finished the lesson successfully.
6. *BatteryOver*: This undesirable State is activated when the smartphone is switched off because the battery state is over.

The dependability of the ubiquitous learning scenario heavily depends on several contextual contraints, especially the quality of the network connection and the energy provided by the mobile phone's battery. In fact, a smartphone is, in general, limited and for sure not keeping pace as the mobile devices are crammed up with new functionalities [19]. For this reason, designers elaborate this technical study.

Figure 5 summarizes this study and shows values of power consumption of different activities that learners had to realize in the learning session. The power consumption is quite higher when student download lesson, and drop into the elearning client all widgets needed for providing video, son and representation.

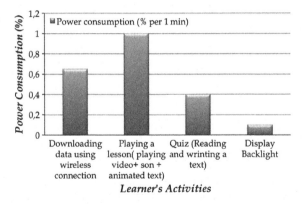

Fig. 5 Energy consumption on a smartphone when switched on

6.2.2 Simulation and Formal Verification

Based on automata presented above, tracks simulations are generated visualizing all possible interactions between different actors. A screen dump of the simulation of the designed educational scenario is below (see Fig. 6). In order to help designers to improve their educational scenario and to obtain better outcomes, through the generated simulations, we try to localise design errors, to answer and to verify the following questions:

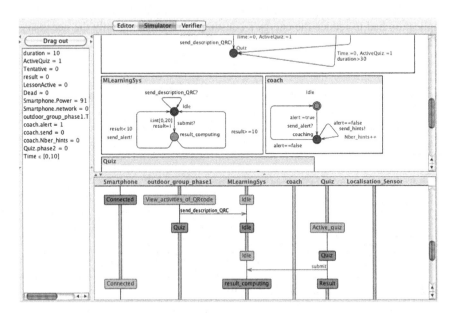

Fig. 6 Simulation of the modeled learning scenario

Table 4 The simulation's parameters

Parameter	Value
Duration of the simulation	240 units of time
Number of learners	40
Phone's energy	100%

- *Does the description of the learning scenario clearly define time constraints for each activity and the whole scenario?*
- *Is there any situation of deadlocks, liveness or starvation?*

We fixe different parameters of the learning scenario's simulation as shown in Table 4. A possible situation of deadlock is detected within tracks simulation of the whole scenario; In fact, students could have a period of inactivity especially when they begin their lesson in the first zone (EDGE) where the network connectivity is lower and smartphone' energy is limited. To check this property, we are based on reachability property that is considered as the simplest form of properties. We ask whether a given state formula, possibly can be satisfied by any reachable state. Another way of stating this is: **Does there exist a path starting at the initial state, such that** *"the battery state"* **is eventually over along that path?**

We traduce this property in temporal logic formula: "**E** <> *Smartphone. BatteryOver*" and this property is verified as shows Fig. 7.

This experiment shows that 37% of students could be blocked at this stage (see Fig. 8). They couldn't progress in the lesson because the state of energy is low. Then, we conclude that dependability is not assured and we had to adapt mobile application to the learning environment's context (if the student is located in *Zone 1*, the mobile application had to reduce the energy's consumption by loading only the necessary widgets.

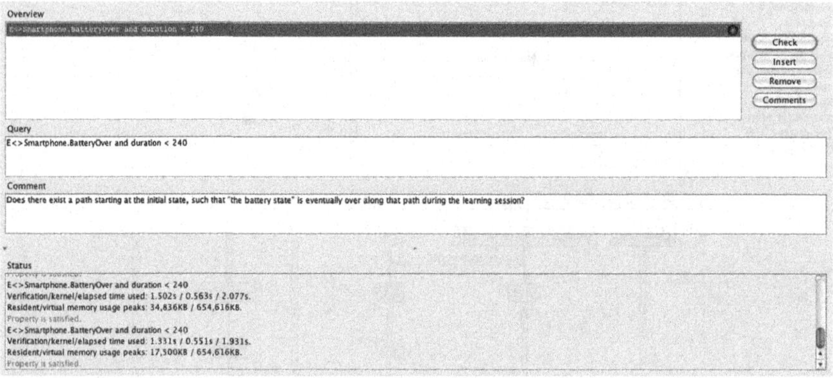

Fig. 7 Checking the reachability of the battery state

Proportion of Learners

■ deadlock (Battery is over)   Lesson finish successful

Fig. 8 The experiment's outcomes

6.3 Results and Discussion

The final evaluation was conducted using interviews with five pedagogical designers. During the interview, the engineers have addressed several points about there experiences with the presented methodology. Hereafter a summary of the main points that have been discussed:

- Fours engineers of five have found the methodology very useful in identifying the inconsistencies of functional aspects of the learning scenario. In fact the culture of classical learning scenario evaluation does not highlight the importance of these functional aspects (like time, context of learning,...). On the one hand, thanks to the formal modelisation, we can simulate and check all possible execution exhaustively. On the other hand, it allows us to redefine the conceived learning scenario deeply and at early stage in order to avoid the problems highlighted by the evaluation (or on the contrary, to rethink the first design).
- Two engineers have already some experience in developing large learning scenario. So they were aware of the cost of their implementation and revision. Consequently, they pointed out that the early evaluation stage performed before implementation and deployment has helped to revise some fundamental decision without having to conduct costly implementation. In fact, the contextual and formal evaluation shows us if the resulting evaluation is exactly what we expect from the modelisation of learning scenario (and from the learner) before the implementation stage. It may happen, for example, that we want to create a correctly and general design for a complex scenario but the result of the formal evaluation shows that we lack in contextual constraints definition and specification.
- Three engineers of five have pointed out that the formal modeling and verification is not so easy to build. Indeed, this formalism is not very well known by all conceptual designers. For this reason, we are developing a formal modelisation assistant tool to help no-skilled designers to evaluate the conceived learning scenario.

7 Conclusion

In this paper, we presented different dimensions that could be evaluated in learning scenario. We proposed a formal framework evaluation to support a contextual evaluation in dependable learning environment. This framework based on formal modelisation and verification allows to designers to simulate the conceived learning scenario with contextual constraints.

Complex content management is employed in m-learning, when there is an interaction between teachers and students. The limit is established by the available technology. It is still a determinant factor in such technologies, the transmission cost and speed, which affects the communication logistic between teacher and student and has a direct repercussion on the learning type. For this reason, we have attempted to demonstrate the benefits of the proposed approach by presenting a realistic case study. We think that this formal methodology of evaluation provides useful information for the re-use and the reengineering of learning materials. In the one hand, it can solve many difficulties in getting proper information about the students and their behaviour (ethical problems, tracks' collect,...).

In the other hand, the use of a rigorous formalism to describe a model of deployment allows a better and more precise understanding of the planned scenario. It provides the designer with an analytical model that helps him to detect errors and learner's difficulties through test case generation and deductive verification. Finally, it creates an environment that simulates an external reality and learner's behaviour.

For going farther this first result, we are currently working on two directions: Firstly, we will attempt to deepen our proposal and to apply formal verification relying on temporal logical formulas and model checking engine. Secondly, in order to assist the designer in expressing these properties, we are developing a tool that helps him/her to draw this business requirement and to generate error report automatically.

References

1. Magal-Royo, T., Peris-Fajarnes, G., Montañana, I.T., Garcia, B.D.: Evaluation methods on usability of m-learning environments. Interact. Mob. Technol. 1(1), 22 (2007)
2. Lê, Q., Lê, T.: Evaluation of educational software: theory into practice. Technol. Teach. pp. 115–124 (2007)
3. Laprie, J.C.: Dependable computing and fault-tolerance. Digest of Papers FTCS-15. pp. 2–11 (1985)
4. Avizienis, A., Laprie, J.C., Randell, B., Landwehr, C.: Basic concepts and taxonomy of dependable and secure computing. IEEE Trans. Dependable Secure Comput. 1(1), 11–33 (2004)
5. Yang, S.J.: Context aware ubiquitous learning environments for peer-to-peer collaborative learning. J. Educ. Technol. Soc. 9(1), 188–201 (2006)
6. Yahya, S., Ahmad, E.A., Jalil, K.A., Mara, U.T.: The definition and characteristics of ubiquitous learning: a discussion. Int. J. Educ. Dev. Using Inf. Commun. Technol. (IJEDICT, Citeseer (2010)
7. Lyytinen, K., Yoo, Y.: Ubiquitous computing. Commun. ACM 45(12), 63 (2002)

8. Henricksen, K., Indulska, J., Rakotonirainy, A.: Infrastructure for pervasive computing: challenges. In: GI Jahrestagung vol. 1, 214–222 (2001)
9. Elissavet, G., Economides, A.A.: An evaluation instrument for hypermedia courseware. Educ. Technol. Soc. **6**(2), 31–44 (2003)
10. Licea, G., Favela, J., García, J.A., Aguilar, J.A.: A pattern system supporting qos for synchronous collaborative systems. In: Proceedings of the IEEE Conference on Protocols for Multimedia Systems-Multimedia Networking. IEEE, pp. 223–228 (1997)
11. King, F.B., Mayall, H.J.: Asynchronous distributed problem-based learning. In: Proceedings of the IEEE International Conference on Advanced Learning Technologies. IEEE, pp. 157–159 (2001)
12. Nielsen, J., Molich, R.: Heuristic evaluation of user interfaces. In: Proceedings of the SIGCHI Conference on Human Factors in Computing Systems, ACM, New York, 249–256 (1990)
13. Polson, L.C., Wharton, P.C., Rieman, J.: Testing a walkthrough methodology for theory-based design of walk-up and use-interface. Proc. ACM CHI. **90**, 235–242 (1990)
14. Zorrilla, M.: Methods and Supporting Technologies for Data Analysis. Data warehouse technology for e-learning, pp. 1–20. Springer, Berlin (2009)
15. Ben Sassi, M., Laroussi, M.: The engineering of tracks for the standard ims ld. In: Proceedings of the IEEE International Conference on Education and E-Learning Innovations ICEELI' 2012, sousse, Tunisia (2012)
16. Jia, Z., Han, X.: Construction of evaluation system on multimedia educational software. (2013)
17. Coronato, A., De Pietro, G.: Formal design of ambient intelligence applications. Computer **43**(12), 60–68 (2010)
18. Bengtsson, J., Larsen, K., Larsson, F., Pettersson, P., Yi, W.: UPPAAL a Tool Suite for Automatic Verification of Real-Time Systems. Springer, Heidelberg (1996)
19. Perrucci, G.P., Fitzek, F.H., Widmer, J.: Survey on energy consumption entities on the smartphone platform. In: Proceedings of the IEEE 73rd Vehicular Technology Conference (VTC Spring), IEEE 2011, 1–6 (2011)

Improving Content Recommendation in Social Streams via Interest Model

Junjie Zhang and Yongmei Lei

Abstract The current microblog recommendation approaches mainly consider users' interests. But because user's interests are changing dynamically and they have low activity, it's hard to build user interest model. In this paper, we propose a new approach to recommend information based on multiaspect similarities of interest and new dynamic strategy for defining long-term and short-term interests according to user's interest changing. Recommended information is ranked by two factors: the similarity between user's interest and information, tie-strength of user interest. We implemented three recommendation engines based on Sina Microblog and deployed them online to gather feedback from real users. Experimental results show that this method can recommend interesting information to users and improve the precision and stability of personalized information recommendation by 30%.

Keywords Recommender system · Naive bayes · Interest model · Microblog

1 Introduction

Information overload has appeared in many places, such as websites, e-commerce and so on, but social network has solvedthe problem of information overload for us to some extent. We read information, which send by our friends, so we avoid much information that we are not interested. But when we follow more people, we will get more information, we should spend more time to choose what we are interested. In the end, information overload appears in social network.

J. Zhang (✉) · Y. Lei
Computer Engineering and Science, Shanghai University, Shanghai, China
e-mail: jadsin@shu.edu.cn

Y. Lei
e-mail: lei@shu.edu.cn

© Springer International Publishing Switzerland 2015
R. Lee (ed.), *Computer and Information Science*, Studies in Computational
Intelligence 566, DOI 10.1007/978-3-319-10509-3_5

Recommender can solve information overload problem effectively. The most famous method is collaborative filtering (CF), which can recommend valuable information according users' preferences. This approach can be divided into different types according to different directions, such as collaborative filtering based on items [9], collaborative filtering based on users, but they all require users to rate items and record their purchase history.

However social network is different, they have no items to rate, although users can rate every story, but this will lose much information contained in these stories. For recommending micro-blog, researchers are more like to use other information such as textual content [1, 17] and LDA topic model [10] etc.

There are many recommendation researches based on micro-blog, Yu et al. [19] transmit relationships into graph, then recommend friends based on mixed graph. Weng et al. [18] researched how to find topic-sensitive influential twitterers; Chen J regarded three factors for recommending conversations in online social streams [4]. In another paper [5], Chen J introduced URL recommendation on Twitter, they used three properties: recency of content, explicit interaction among users and user-generated content.

Recent researches research more about recommending friends, conversations and so on, they referred to comments, tags or replays. However, there are few researches concerning content recommendation, and not all people in social network are interested in all the information. To save time for user to get interesting information, social networks provide effective methods, Facebook used EdgeRank algorithm [8], particularly favoring recent and long conversations related to close friends, to recommend social stream. Instead, Twitter adopts simple rules to filter streams. Sina micro-blog has a function called intelligent ranking, which can recommend information, but if users don't use the system usually, its performance begins to decline. However, we don't know how they realized.

It's hard to find interesting information for user because of four difficulties:

- It's difficult to build users' interests. No matter online or offline, user's interests are changing dynamically. Although content can be considered as containing user interest, but it is valuable within a short time since being published.
- Low activity. Unlike recommendation system, people rate items, and then system figures out people's interests according these scores. In social networks, people are more like to read information rather than write or rate, so it's difficult for us to get their explicit information and interests.
- Limit of content's words. In social network, every status is limited within 140 words. To express within limit words, users minimize the usage of repeated words. It's hard to analysis contents using existing models.
- Large amount of data but sparse. Users provide few explicit information in social network, so we use large amount of implicit information, such as information contents, time, tags or followees' interests. However, although there are many followees, most of them can't provide effective interests.

To solve these problems when recommending content, we propose an improved algorithm of building user's interests and measuring the similarity between user and content, and then recommend the most similarity information. We deploy recommendation system using our algorithm online to gather feedback from real users. We asked users to rate for each recommending information produced by different algorithms. The result scores can help us to compare the performance of different algorithms.

The rest of the paper is structured as follows. First, we discuss how existing research relates to our work. We then propose an improved algorithm and detail our design of algorithm and deploy recommendation system for studying, and then we detail our work with results. We conclude with comparison of our recommendation to others, such as Sina Intelligent Ranking.

2 Related Work

As popular of social network, more and more people use it to get information. But we'll find that we are not interested in every information that posted by our friends, and what's more, because people's preferences are quite different. Paek et al. [12] found that many posts were considered important to one user but worthless to another user, that is personal preferences is important for content recommendation. In social networks, there is little explicit information to build user's interests, we can't rate items to indicate our preferences. But statuses contents that posted by us can represent. We can extract key words that appear many time in our posted contents to indicate our interests.

There are many recommendations based on textual content, such as websites [13] and books [11]. For example, to recommend websites, Pazzani et al. [13] created bag-of-word profiles for individuals from their activities, and chose websites that are most relevant to the profile as recommendations. Because activities of an individual are often insufficient for creating useful profiles, Balabanovic et al. created profile from a group of related individuals [1].

Since the fast update rate of micro-blog and users' interests are changing dynamically, we divide users' interests into long-term interests and short-term interests. To improve the precision of recommendation, we consider three factors to build users' interests:

- Tags: People can use tags to represent their interests. We can also find users that has common interests through tags.
- Contents: When users send a status, they may want to transmit information or express their status, but these all contains their interests. We use the contents of the latest one month to build user's short-term interest.
- Followees: In social networks, most people use it to get information, they rarely write news, those types of people have low activities. We follow someone to get useful information from them, we are interested in what followees post and have some common interests with them, so when constructing interests, we consider

using their followees' interests to extend their interests, especially when users have low activities. Prior work [16] had proved the effectiveness of combining text content from a group to capture the interest of single user.

As mentioned earlier, user's contents are interesting within a period of time since being published. Boyd, D mainly researched the conversational usage of retweets in Twitter, he thought the time that tweets are posted was important [3]. When considering contents to build interests, we also consider the influence of time.

In social network, the importance of individuals is different. For example, my friend David, he always read information from sina micro-blog, but he posts few, so he is at a normal position , once he post a status, we may think his information is not very valuable. And my friend Ange, she reads information as well as writes always, we always get information from her posts, so we think her news or interests are more important than David to us. So we divided followees, when using their interests to construct user's interests, we attach different importance to different followees.

Tie-strength is a characterization of social relationships between people [7]. Gilbert et al. [6] suggested that tie-strength may be a useful factor for filtering messages in Facebook news feeds and measured tie-strength with many dimensions. Chen et al. [4] was inspired by Gilbert, they used three dimensions to recommend conversions: existence of direct communications, frequency of such direct communications, tie-strength between the two and their mutual friends and Chen J et al. Measured tie-strength between users according to the frequency of communications among users and tie-strength between the two and their mutual friends.

People follow someone because of their common interests, so we follow common friends indirectly indicating that we have common interests. Inspired by Chen, when measuring tie-strength between users, besides considering the similarity between their owns' interests, we also consider their common friends' tie-strength of interests.

When we want to recommend information to users, we calculate the similarity between user's interests and content and tie-strength between users based on interests, and then recommend high scores information.

3 Multiaspect Similarity of Interest

To find interesting information for users in social streams, we propose a new method to recommend information based on multiaspect similarities of interest, which considers the similarity between user's interests and information and tie-strength between users' interest. Due to dynamical changing of interest, we use their long-term and short-term interests to build users' interest model, and we don't merely use total number of common friends to measure tie-strength, we also take their common friends' interests into account, which will improve the precision in social network based on interests. We recommend the most similarity information to users according their interests. Next we will introduce our method through three parts.

3.1 *Interest Model*

Due to the dynamical changing of user's interests, we divided interests into long-term and short-term interests, which can build user interests more precise.

Users' short-term interests show change of his recent statuses. In our research, we followed the approach in Pazzani et al. [13], we build a bag-of-words profile for each user to represent his short-term interests. Unlike in Pazzani et al., where the profile consists of words from web pages that the user has rated explicitly, we build the profiles from users' recent one month micro-blog contents with TF-IDF model [15].

However, since each status contains less than 140 words, the constructed bag-of-words had lower weight, so we use TweeTopic technique [2] to enrich the content of each status. We feed each micro-blog to Google CSE [14] search engine, and extract the returned documents to get key words to represent this information's vector.

Users are more interested in recent news, and the value of information begin to decay with time passing, so to predict users' interests better, we modified TF-IDF model, and added time penalty function as (1)

$$f(u) = \sum_{n=1}^{N} \sum_{x} (TF_x {}^* IDF_x {}^* T_n) \tag{1}$$

where N is total number of user's micro-blogs, x is words in document n. TF_x is the frequency of word x in micro-blog n; IDF_x is log(total number of micro-blogs / total number of micro-blogs that contain word x). T_n is time penalty function, there are many functions, we use one of them as (2)

$$T = \frac{1}{1 + \alpha |T_0 - T_i|} \tag{2}$$

where T0 represents current time, Ti represents time of micro-blog i that be posted, α is weight. When α is assigned to different value, the decay speed is different. Experiment shows that when α is assigned to 0.5, the decay speed is better.

Unlike short-term interests, users' long-term interests don't change usually, but it can increase. User posts a status, which is valuable within a short time, and when time is passing, the value of this status is decaying, so old statuses only show that users had ever been interested in them, we can't only use contents to build use's interests.

In our research, when building long-term interests, we consider three factors: micro-blog's contents, tags and followees' interests. For most people read information rather than write, using contents and tags merely to build interests maybe not precise. We are inspired by Chen [5], they researched URLs recommendation on Twitter, when building URLs candidate, they considered URLs posted by followees and followees of followees. So we use user's followees' interests to enrich their interests.

However, we have analyzed the recent one month contents for constructing short-term interests. So to build long-term interests, we don't need to analyze these contents any more. We had built all short-term interests every month, then we can reuse these, but we should add time factor, but this time we use month as unit.

In social relationship, we can't treat every followees as equal, Some people, who have unique views and post many valued information, are extremely important. In our research, we divide followees into two types: Information Source User and Normal User. Information Source Users always provide unique and valued information or unique Opinions on something, most of us like to read news from them, hoping to extend our interests according to their own interests, then we can get more and more interesting and valuable information. But Normal Users are not as import as the other type. They mostly like to read and don't usually express their opinions. So we will attach more importance to Information Source Users than Normal Users.

We use Naive Bayes to divide them, the method includes four characters: number of followees, number of fans, number of bi-followers and number of micro-blogs, the classifier as (3)

$$P(C_i|W) = \frac{P(C_i) \prod_{n=1}^{4} P(W_i|C_i)}{P(W)} \tag{3}$$

where W is consist by four characters, C is the two types.

We build user's followees' interests as follows: for each followee F, if he has been built interest vector, then we use it directly. Otherwise, we use F's all contents that he posted, we then use the modified TF-IDF models to analysis these contents, but we don't feed micro-blogs to search engine for improving efficiency, we get an interest vector as V_F, rank the vector by decreasing order of words' weight in V_F, then we remove these words that have low weight, and choose the top 10% of words to extend user's long-term interests.

During this operation, we gather many words from their followees, and the weight of every word may be bigger. When weight of words becomes bigger, that shows more than one followee are interested in these words, and user is more interested in them. So after gathering all followees' interest vector into one, we set a threshold, only can the weight is larger than the threshold be kept. This will decrease complexity of future computing.

We build user's long-term interests with tags, contents and followees' interests with weight. And we add long-term and short-term interests to construct user's interest vector, because users are more eager to get information, what he pays attention to recently, and we give higher weight to short-term interests than long-term interests. This will improve recommendation's breadth without losing accuracy of recommendation.

3.2 Tie-Strength

Micro-blog's friendship is based on common interests, when users follow the same people, they are more likely to have common interests. If we measure tie-strength by total number of common friends, that lose much information behind relationship, such as whether we really both have common interests with our common friends, whether we are more familiar to user A than user B, besides, that can't make a distinction between information source users and normal users as well.

So in our research, when measuring tie-strength between users, we don't use total number of common friends, we consider the tie-strength of their common friends' interests with them and the similarity between users own interests. To distinguish between information source users and normal users, we give higher weight to information source users. The detail of this approach as follow, we assume to calculate tie-strength between A and B in Fig. 1.

In Fig. 1, C, D and E are common friends of user A and user B, C and E are information source users, they have higher weight as a in (4), D is normal user, he has low weight. α_1 is the tie-strength between A and C, others define as same. We suppose A and C's common interests are $<V_1, V_2, ..., V_n>$, A's interest vector mapping in common interests with weight is $<W_1, W_2, ..., W_n>$, and is called V_a, and C is $<W'_1, W'_2, ..., W'_n>$, called V_c, then we use cosine similarity to calculate α_1

$$\alpha_1 = \cos(V_a, \partial V_c) \quad 0.5 < a < 1.0 \tag{4}$$

where a is weight of source users. D is normal user, his weight is 1-a. So the function of calculating similarity between A and D as (4), but the weight factor a is replaced by 1-a.

But when calculating A and B's similarity of interests, we use cosine similarity without weight, because them are equal, and then we build tie-strength through their owns and common friends as (5)

Fig. 1 User's relationship model with common friends

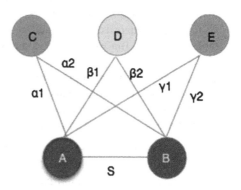

$$Ti(A, B) = S + \sum_{i=1}^{n} \alpha_1 \alpha_2 \tag{5}$$

where α_1 is tie-strength between user i and A, α_2 is tie-strength between user i and B, n is total number of A and B's common friends.

3.3 Calculate Scores

At last we calculate the similarity between user and news as part of final score, using given user's interests vector and topic vector of recommending content, and tie-strength between users as another part of final score, we use weight α to adjust combination of two parts as (6). We recommend higher scores information to user.

$$Score(u, i) = \partial \cos(V_u, C_i) + (1 - \partial)Ti(u, B) \tag{6}$$

This calculates how user u is interested in new i. V_u is user u's interest vector, C_i is a set words of new i, Ti(u,B) is tie-strength between user u and B, which user B is the owner of new i.

4 Dataset and Experiment

In our research, to improve the precision of recommending, we use followees' interests to extend user's interests, we also use relationships between followees to measure tie-strength, so we compare the following five ranking algorithms for research:

- Default: When we use social network, we read from social streams, which contains information that posted by our followees. It's ranked by time that information is posted. We use this recommendation as baseline for other algorithms.
- Intelligent Ranking: Sina micro-blog has a recommendation system, which called Intelligent Ranking, it can rank information according users' interests.
- No followees' interests: When building user's interests, we don't use their followees' interests. But we use their common friends tie-strength to measure their tie-strength.
- No tie-strength: Building user's interest model, we use their followees' interests, but we don't use common friends' tie-strength to measure their tie-strength.
- With followees' interests and tie-strength: When building user's interests, we use their followees' interests. And we also consider tie-strength when recommending.

In our recommendation system, we should build users' interest model firstly. We use three factors: contents, tags and followees' interests. When users authorize to use our system, we get these information and construct their interests using above algorithms. Then we update their interests incrementally, we just need to analyze new parts, such as new followees and new statuses. We put this operation offline at midnight every day. When information comes, we have had users' interests already, we compute the similarity between user's interests and topic vector of information and tie-strength between user and owner of this information, and each information gets a score, we rank information according to these scores and recommend higher scores information.

When adding followees' interests to user's interest model, we need to distinguish followees. Different type of followees has different weight. We experiment Naive Bayes to make sure it can help us to divide them. We get 150 users' data from sina APIs, each user's data contains four elements: number of followees, number of fans, number of bi-followers, number of statuses, and choose 130 users randomly to compose train set, and rest 20 users compose test set. We use the train set to train the classifier, and get percentage of influence of each property, and then we use test set to test the accuracy of classifier. We repeated this operation ten times, and figure out average error rate, and we use the average percentage of each property in our recommendation system.

When constructing user's interests, we separate all contents into groups since users post first status, each group contains one month's contents. Short-term interests only use the last group, long-term interests use all contents exclude last month. So we can reuse short-term interests rather than analyzing all contents repeatedly. In particular, user's followees may be too many, which we are not able to process all of them. Sarwar et al. [16] have shown that by considering only a small neighborhood of people around the end user, we can reduce the set of items to consider, we'll remove some persons, who have statuses less than 20, this will reduce the amount of dataset effectively.

We deployed recommendations online for two weeks, and invited 52 active users to participate in our research. We divided these users into three groups, group one(G1) will use our system every day, group two(G2) use system every other day, group three(G3) use system every three day. Every time users came, we achieved first one hundred news in Sina Intelligent Ranking, and pick up fifty news from beginning, middle and end position, then we ask users to rate every new with five star according to their interests, which like Fig. 2. And we feed these fifty news to another four algorithms to rank them, and the five algorithms generate a total of 250 recommendations each time, every new has a score in each algorithm.

Then we use ASPT(average scores per time) to estimate the accuracy of each algorithm. Method of calculating ASPT as (7)

$$ASPT = \frac{1}{N} * \frac{\sum_{i=1}^{N} \sum_{j=1}^{M} S_j P_j}{\sum_{i=1}^{N} S_{\max}} \qquad (7)$$

where N is total numbers of groups, M is total numbers of each group, here we choose 50; S_j is score of new i that users rate. In each recommendation, new i is in P_i position. $S_{\max}$ is the max score that one micro-blog can get , we define it as five. If algorithm has higher ASPT, users are more satisfying with this algorithm.

Fig. 2 Online test website layout

5 Discuss

We ran our study online for two weeks and invited 52 real users to participate, each user produced one group score data each time of each algorithm, users in first group will totally produce 1,190 groups score data; users in second group will totally produce 595 groups score data; users in third group will totally produce 360 groups score data, and each group contains 50 scores, because each time we recommend 50 news.

Besides we have experiments for dividing followees types. We use 150 users' data for experiments for training and testing the classifier, and we repeat training for ten times, we get the influence of each property for different types of users.

In the following part, we will discuss our experiments and results from the followees division, user's interest model, tie-strength and stability of algorithms.

5.1 Followees Division

Figures 3 and 4 are the results of test of Naive Bayes. In each figure, every column represents one experiment, and x- axis is error rate of each test, y-axis is percentage of influence of each property.

Figure 3 represents information source users. From this figure, bi-followers and friends have lower influence, but fans and statuses hold higher percentage. Information source users always post news, many users follow them to get newer and

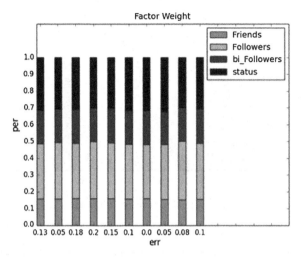

Fig. 3 Error rates of source users and influence percentage of each property

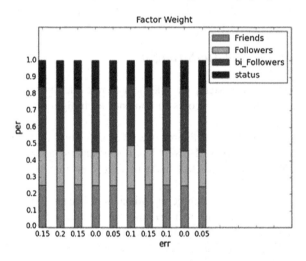

Fig. 4 Error rates of normal users and influence percentage of each property

valuable information. We found that bi-followers are mostly their followees, that's to say their followees mostly follow them back, because the information source users are more like to follow each other, so they can get more value information from each other.

Figure 4 represents normal users. From figure, fans and statuses don't get high percentage, especially statuses, because this type of users likes to read news, not to write, they want to get more information from others, so followees get a higher percentage. But bi-followers has highest percentage of influence, unlikely source users, normal users' bi-followers are mostly their fans.

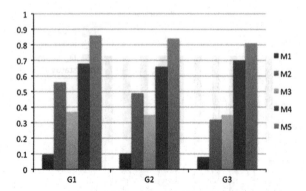

Fig. 5 Distribution of scores of each algorithm rated by each group

So we can use the four characters above to divide users' type, because information source users have more statuses and fans, and bi-followers are mostly their followees; normal users have little statuses and fans, and bi-followers are mostly their fans.

5.2 Interest Model and Tie-Strength

When volunteers use our system, we ask them to rate each news according their interests. At last we use these scores to compare each algorithm. The results as Fig. 5.

To reduce the complexness of figures, we give every algorithm an abbreviation. We call the default algorithm as M1; Sina Intelligent Ranking as M2; No followees' interests as M3; No tie-strength as M4; and with followees' interests and tie-strength as M5.

From Fig. 5, M1 has lowest degree of satisfaction, because M1 ranks by time, so user should make much time to filter information; the others consider user's interests and contents, they get higher scores, and our recommendation is better. As we mentioned earlier, tags and contents can't always build our interest model precisely due to low activity, so M3's performance is lower than Intelligent Ranking. When considering followees' interests, M4 and M5 have higher precision, because through their followees' interests, we extend users' interests. If we consider tie-strength between users, the satisfaction of recommending improved about 12 %. So tie-strength is an important factor for recommending.

5.3 Stability of Algorithms

From Fig. 6, When user don't get in website for a long time, Intelligent Ranking performance begins to decline, We find that Sina Intelligent Ranking recommends information that is mostly posted by who we always prefer to read from, it mays

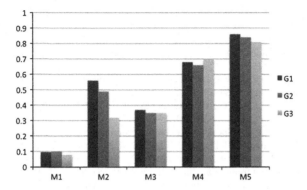

Fig. 6 Distribution of scores from another view

record these persons, who we always read news from, not our interests, so it's not precise when we rarely use it.

However, we construct users' interests from long-term and short-term interests, through this way, users' interest model won't change rapidly, so our algorithms are more stable.

So When recommending news, we consider users' long-term and short-term interests to improve the stability of algorithm, and construct users' interests with their followees' interests to improve the precision of recommending.

6 Conclusion and Future Work

In this paper, we have studied content recommendation on sina micro-blog. We implemented three algorithms. We compared each algorithm's performance with Sina' Intelligent Ranking algorithm through the usage results of 52 sina users, we found that our algorithm, which consider user's long-term and short-term interests and tie-strength between users, are more precise and stable than Intelligent Ranking algorithm.

As mentioned above, there is more implicit information than explicit. In our research, we have used users' behavior of following, content and tags. Future research may add more dimensions to construct user's interests and tie-strength. People use micro-blog for different purpose, such as information purpose and social purpose [4], but in our research we don't distinguish these too much. We'll pay more attention to recommendation for different usage purposes in our future work.

In our ranking algorithm, although we consider the importance of individuals is different, in social network, there are many areas of interest, but we regard every user from different areas of interest as same. In future research, we will refine grain size, and divide people into more different groups according their interests, and choose the highest words from each group.

References

1. Balabanovic, M., Shoham, Y.: Fab: content-based, collaborative recommendation. Commun. ACM **40**(3), 66–72 (1997)
2. Bernstein, M., Suh, B., Hong, L.: Eddi: Interactive topic-based browsing of social status streams. In: Proceedings UIST'10. 2010
3. Boyd, D., Golder, S., Lotan, G.: Tweet tweet retweet: Conversational aspects of retweeting on Twitter. In: Proceedings HICSSS'10, 2010
4. Chen, J., Nairn, R., Chi, E.: Speak little and well: recommending conversations in online Social streams. In: Proceedings CHI'11. pp. 217–226. ACM, New York, USA (2011)
5. Chen, J., Nairn, R., Nelson, L., Bernstein, M., Chi, E.H.: Short and tweet: experiments on recommending content from information streams. In: Proceedings CHI'10, ACM Press (2010)
6. Gilbert, E., Karrahalios, K.: Predicting tie strength with social media. In: Proceedings CHI'09. 2009
7. Granovetter, M.S.: The strength of weak ties. Am. J. Sociol. **78**(6), 1360–1380 (1973)
8. Kincaid, J.: EdgeRank: The secret sauce that makes Facebook's news feed tick. Retrieved from http://techcrunch.com/2010/04/22/facebook-edgerank 2010
9. Linden, G., Smith, B., York, J.: Amazon.com recommendations: item-to-item collaborative filtering. IEEE Comput. Soc. **7**(1), 76–80 (2003)
10. Mehrotra, R., Sanner, S., Buntine, W., Xie, L.: Improving LDA topic models for Microblogs via tweet pooling and automatic labeling
11. Mooney, R.J., Roy, L.: Content-based book recommending using learning for text categorization. In: Proceedings of ACM DL'00. pp. 195–204 (2000)
12. Paek, T., Gamon, M., Counts, S., Chickering, D.M., Dhesi, A.: Predicting the importance of newsfeed posts and social network friends. In: Proceedings AAAI'10. 2010
13. Pazzani, M.J., Muramatsu, J., Billsus, D.: Syskill & webert: Identifying interesting web sites. pp. 54–61, AAAI/IAAI (1996)
14. Sahami, M., Heilman, T.D.: A Web-based kernel function for measuring the similarity of short text snippets. In: Proceedings of WWW'06. 2006
15. Salton, G., Buckley, C.: Term-weighting approaches in automatic text retrieval. Inf. Process Manage. **24**(5), 513–523 (1988)
16. Sarwar, B.M., Karypis, G., Konstan, J.A., Riedl, J.: Recommender systems for large-scale E-Commerce: Scalable neighborhood formation using clustering. In: Proceedings of ICCIT'02. 2002
17. Schein, A.I., Popescul, A., Ungar, L.H., Pennock, D.M.: Methods and metrics for cold-start recommendations. In: proceedings of SIGIR'02. 2002
18. Weng, J., Lim, E., Jiang, J., He, Q.: TwitterRank: finding topic- sensitive influential twitterers. Proceedings of the 3rd ACM international conference on web search and data mining (WSDM □ 10), pp. 261–270. ACM, New York, USA (2010)
19. Yu, Y., Qiu, G.H., Chen, A.P.: Friend recommendation algorithm based on mixed graph in online social networks. New technology of library and information service. 2011

Performance Evaluation of Unsupervised Learning Techniques for Intrusion Detection in Mobile Ad Hoc Networks

Binh Hy Dang and Wei Li

Abstract Mobile ad hoc network (MANET) is vulnerable to numerous attacks due to its intrinsic characteristics such as the lack of fixed infrastructure, limited bandwidth and battery power, and dynamic topology. Recently, several unsupervised machine-learning detection techniques have been proposed for anomaly detection in MANETs. As the number of these detection techniques continues to grow, there is a lack of evidence to support the use of one unsupervised detection algorithm over the others. In this paper, we demonstrate a research effort to evaluate the effectiveness and efficiency of different unsupervised detection techniques. Different types of experiments were conducted, with each experiment involves different parameters such as number of nodes, speed, pause time, among others. The results indicate that K-means and C-means deliver the best performance overall. On the other hand, K-means requires the least resource usage while C-means requires the most resource usage among all algorithms being evaluated. The proposed evaluation methodology provides empirical evidence on the choice of unsupervised learning algorithms, and could shed light on the future development of novel intrusion detection techniques for MANETs.

Keywords Mobile ad hoc networks (MANETs) · Anomaly detection · Unsupervised learning

1 Introduction

Mobile Ad Hoc Network (MANET) is a self-configuring network in which each mobile device or node can communicate with every other device or node via wireless links [3]. Unlike cellular wireless networks that depend on centralized access

---

B.H. Dang · W. Li (✉)
Graduate School of Computer and Information Sciences,
Nova Southeastern University, Fort Lauderdale, FL 33314, USA
e-mail: wei@nova.edu

B.H. Dang
e-mail: dbinh@nova.edu

© Springer International Publishing Switzerland 2015   71
R. Lee (ed.), *Computer and Information Science*, Studies in Computational
Intelligence 566, DOI 10.1007/978-3-319-10509-3_6

points (such as routers, gateways, etc.), MANETs lack a fixed infrastructure. Nodes in MANETs cooperate with each other by relaying packets to each other when they are within a certain radio transmission range. In MANETs, nodes from different geographic areas may leave and join the network at any time, and each mobile node requires battery power to perform its tasks. The lack of a fixed infrastructure, limited resources (e.g., CPU, memory, battery power, etc.), and frequent changes in the network topology present numerous challenges in MANET security.

Intrusion detection systems (IDS) intend to provide efficient mechanisms to identify attacks in MANETs. There are generally three IDS approaches in the research literature: misuse based, anomaly based, and specification-based. Misuse based detection techniques are very accurate, but they impose a serious limitation on intrusion detection since knowledge of the attacks must be known in advance. Given the characteristics of MANETs, these techniques are expensive and not able to detect novel attacks [5]. Specification-based detection techniques use system or the routing protocol behavioral specifications to detect attacks. However, building specified behaviors of the system can be time-consuming. Recent research efforts have been focused on anomaly detection techniques. These anomaly detection techniques include supervised-based and unsupervised-based machine-learning methods. Although many anomaly detection techniques have been proposed in the literature, a comprehensive comparative study of unsupervised intrusion detection techniques has not yet been performed. Different methodologies were tested on a variety of platforms with different simulation datasets. As a result, there is no clear evidence to support the use of one unsupervised detection algorithm over the others. Even if a decision on the detection algorithm can be made, there is no guidance how to configure the algorithm to achieve better performance.

In this paper, we demonstrate part of an ongoing research effort on the comprehensive evaluation of unsupervised learning algorithms, more specifically, the clustering algorithms, the Principal Component Analysis (PCA), and the Support Vector Machine (SVM). These algorithms are popular unsupervised learning algorithms, and have been applied in previous research efforts. In our research, we used a well-defined experiment methodology so that different algorithms can be compared objectively. Four types of attacks on MANETs were simulated and the unsupervised learning intrusion detection techniques were applied. The intrusion detection shows different levels of success, which will be presented later in this paper.

The remainder of this paper is organized as follows. Section 2 reviews recent research work that is closely related to our research. Section 3 provides an overview of the AODV protocol and potential attacks against it. Section 4 describes a well-defined methodology to conduct the experiments. Section 5 presents the experimental results and analysis on these results. Section 6 shows lessons learned in this research and indicates possible directions for future work.

2 Related Work

Different machine learning techniques have been applied to the intrusion detection problem in MANETs. The first major category is based on supervised learning, in which the training dataset is labeled. A variety of approaches have been proposed. For example, Mitrokotsa et al. [10] evaluated the performance of intrusion detection techniques using different classification algorithms such as Multi-Layer Perceptron (MLP), the linear classifier, the Gaussian Mixture Model, the Naïve Bayes classifier, and the Support Vector Machine (SVM). The authors concluded that SVM was the most efficient classifier since its false alarm rate was significantly lower than that of MLP. It was also found that a sampling interval of 15 s was on average the most efficient.

The main drawback of supervised anomaly detection in MANETs is the need for the labeling of data, which could be time consuming if not performed automatically. Sometimes such data is not available especially for a highly dynamic network liked MANETs. To address this issue, several unsupervised learning techniques for anomaly detection in MANETs have been proposed, and some of them are shown as follows. Sun et al. [16] proposed an anomaly detection method in MANETs using a mobility model with different random walk models. Markov Chain was then applied to calculate the alert signal of recent routing activities. Li and Joshi [9] applied the Dempster-Shafer theory to reflect the uncertainty or lack of complete information and to combine observation results from multiple nodes in the context of distributed intrusion detection. Avram et al. [4] analyzed deceiving attack in AODV and selective forward attack and black hole attack, a type of node isolation attack, using self- organizing maps (SOMs). Statistical regularities from the input data vectors were extracted and encoded into the weights without supervision. Nakayama et al. [11] proposed an anomaly intrusion detection approach using features based on characteristics of routing disruption and resource consumption attacks. In the proposed method, the authors applied a dynamic learning process to update normal network profile using weighted coefficients and a forgetting curve. Similarly, Alikhany and Abadi [2] applied a weighted fixed width clustering algorithm. The authors concluded that both the detection rate and false alarm rate were improved significantly when using an adaptive normal profile.

These approaches have shown varied degree of success in MANET intrusion detection techniques and some are quite promising; however there is a need for a comprehensive empirical comparison of various unsupervised learning methods on the same platform.

3 Attacks on the AODV Routing Protocol

3.1 An Overview of AODV Routing Protocol

Ad hoc On-Demand Distance Vector (AODV) Routing protocol is a core protocol for MANETs and other ad-hoc networks, and was defined in RFC 3561 [13]. ADOV is an on-demand protocol that finds a route by flooding the network with requests. It mainly includes a route discovery process and a route maintenance process. In a route discovery process, a source node is a node that initiates the route discovery process when it needs a route to the destination. A source node makes a request by broadcasting a Route Request (RREQ) packet to its neighboring nodes. If the neighboring node is not the intended destination, it will forward the packet to the next immediate nodes also known as intermediate, or a forwarding node, and the process continues until it reaches the destination. When a node determines that it has a route to the destination, it will send a Route Reply (RREP) packet, which contains the hop count, life-time, and current sequence number of the destination. When an intermediate node receives and has processed the RREP packet, it will forward the RREP packet towards the source node, which will begin data transmission upon receiving the first RREP packet. In the route maintenance process, if a source moves, a new route discovery process is initiated. If the intermediate nodes or the destination node move, all active neighbors are informed by the Route Error (RERR) message. When a source node receives the RERR message, it will reinitiate the route discovery process [13].

There are other routing protocols for MANETs in the literature, such as Optimized Link State Routing Protocol (RFC 3626), Babel Routing Protocol (RFC 6126), Dynamic Source Routing (RFC 4728), among others. This is an active research area and new protocols are still being proposed. Despite this, our research effort has been focused on the AODV and attacks against it. The research methodology shown in this research can be readily extended to similar or newer routing protocols.

3.2 Classification of Attacks

In the initial experiments of this research, we simulated some typical attacks as shown in the literature [12]. These attacks are shown briefly below. It should be noted that the same methodology can be used to simulate novel attacks on MANETs.

- Route Disruption (RD) Attack: This attack breaks down existing routes or prevents a new route from being established. An example of route disruption attack is the dropping attack.
- Route Invasion (RI) Attack: This attack attempts to insert an attacking node between two communicating nodes. The inserted node is then used to carry out

attacks such as selectively dropping and/or sniffing data packets. The most common attack of this type is the forge reply attack.

- Node Isolation (NI) Attack: This type of attack prevents the target from communicating with other nodes in the network. An example of this type of attack is the blackhole attack.

- Resource Consumption (RC) Attack: This attack wastes resources of the victim. For example, a malicious node may increase the RREQ ID field in the IP header, and send an excess number of RREQ messages. As a result, the network will become congested with a large number of forged RREQ messages, thus wasting the network resources. Typical attacks of this type are the denial of service attack and flooding attack.

4 Unsupervised Learning Techniques

4.1 Clustering Techniques

The clustering techniques applied in the experiments include popular algorithms such as K-means, C-means, and Fix-Width (FW) clustering. K-means algorithm is a type of partitional clustering algorithm, and it is one of the most efficient clustering algorithms in terms of execution time when compared against several other clustering algorithms. Given some data, K-means algorithm partitions the points in k groups such that the squared error between the empirical mean of a cluster and the points in the cluster is minimized. It is well known that the problem of finding an optimal value for k is NP-hard [8], and in many cases it is determined through experience. In our experiments, for the initial value, we chose $k = 2$ assuming that normal and anomalous traffic in the training data will form two different clusters.

C-means algorithm is a fuzzy clustering algorithm in which one record is allowed to belong to two or more clusters. C-means algorithm is similar to K-means except that the membership of each point is defined based on a fuzzy objective function, and a point may not necessarily belong to one cluster.

FW is a clustering technique in which sparse clusters are considered as anomalous based on a given width of the clusters. However, the implementation of the method may significantly diverge on how the cluster width and how the percentage of the largest clusters is determined [1].

4.2 Principal Component Analysis

PCA is a popular unsupervised learning technique that coverts possibly correlated variables into linearly uncorrelated variables called principal components. It begins with the calculation of the mean vector and the covariance matrix. The eigenvectors and eigenvalues of the covariance matrix are calculated, and the eigenvectors are rearranged in an order from highest to lowest based on the eigenvalues, so that a subset of the components or eigenvectors will be chosen as the feature vectors based on their order of significance. The most important axis to express the scattering of data is then determined to explore the correlations between each feature [14].

4.3 One-Class SVM

SVM provides a very efficient mechanism for supervised learning. The basic idea of SVM is to find an optimal separating hyperplane, surrounded by the thickest margin, using a set of training data. Prediction is made according to some measure of the distance between the test data and the hyperplane [15]. In our experiments, unlike supervised SVM, unsupervised One-Class SVM (OCSVM) is trained with an unlabeled training dataset, and classes were labeled as positive and negative. In addition, for OCSVM, we assumed that the majority of training data are normal.

5 Experiments

5.1 Features Selection

In our simulation, each mobile node monitored its own traffic and its neighbor's traffic, and used a time slot to record the number of packets according to their types. In the learning process, a time interval was defined, which contains several time slots. The statistical features of interest was calculated from each packet of each trace file of each node during a given time slot. The training dataset was collected during the first time interval. The test dataset was collected immediately after the training interval. Prior research efforts have used features proposed in Table 1 [2, 11]. Similar to these efforts, our research applied 14 traffic features and an AODV characteristic feature.

5.2 Experiment Overview

There were different types of experiments within each of our experiments, which involved different parameters. Each algorithm was applied to the same set of experiments to ensure an objective comparison. Four types of attacks, including the node isolation, route disruption, resource consumption, and route invasion attacks were simulated. Each attack was carried out for the same duration.

Table 1 Types of features

Features	Descriptions
Recv_rreq_ss	Number of received RREQ messages with their own source IP address
Recv_rreq_sd	Number of received RREQ messages with their own destination IP address
Recv_rreq_dsd	Number of received RREQ messages with different source and destination IP addresses
Recv_rrep_ss	Number of received RREP messages with their own source IP address
Recv_rrep_ds	Number of received RREP messages with different source IP address
Frw_rreq	Number of forwarded RREQ messages
Frw_rrep	Number of forwarded RREP messages
Frw_rrer	Number of forwarded RRER messages
Dropped_rreq	Number of dropped RREQ messages
Dropped_rrep	Number of dropped RREP messages
Send_rreq	Number of sent RREQ messages
Send_rrep_sd	Number of sent RREP messages with the same destination address as the node
Send_rrep_ss	The number of sent RREP messages with different destination address as the node
Recv_rrer	Number of received RERR messages
AODV charateristic	Average difference at each time slot between destination sequence number of received RREP packet and stored sequence number in the node

Each experiment started with each node collecting data for itself and from its neighboring nodes. The collected data were preprocessed and features were extracted to identify attacks. In each experiment, we evaluated the performance of each clustering technique with a distance-based heuristic.

We then evaluated the performance of each clustering technique by comparing direct mode against indirect mode. In indirect mode, a normal profile was constructed after the training. During the testing, each data instance was compared against the centroids of all clusters and was assigned to the closest one. In direct mode, we applied the unsupervised detection methods directly to the target dataset. The initial normal profile model obtained from the first interval was chosen for anomaly detection during the testing phase. In addition, different thresholds were tuned to detect whether or not an instance is normal. For K-means and C-means algorithms, we applied the average distance threshold. For FW, we used a width value of 0.5. For PCA, we used the maximum value of the projection distance. For OCSVM, we used radial basis function (RBF, a popular kernel function for SVM) as the kernel function and initialized its gamma parameter to 0.067 (1 per 15 features).

Table 2 Simulation parameters

Run time (seconds)	Example
Number of nodes	30, 100, 200
Mobility model	Random waypoint
Area for 30/other nodes	$1000\,m \times 1000\,m/2000\,m \times 2000\,m$
Max speed (m/s)	5, 10, 20
Pause time (seconds)	10, 50, 100, 200
Max connections	25

5.3 Experiment Setup

Ns-2 (http://www.isi.edu/nsnam/ns/) is a discrete event network simulator, and has been used in a number of research efforts to simulate network traffic. In our research, it was used for each of the experiments. The maximum connection was set to 25, and at a rate of 1 m/s. The traffic load was constant bit rate (cbr) flows with a data packet size of 512 B. The 802.11 Media Access Control layer had a transmission range of 250 m, and a throughput of 2 Mb/s. Specific parameters are shown in Table 2.

We used LIBSVM library [6], Cluslib-3.141 tools [7] to carry out experiments with OCSVM, K-Means and C-Means, respectively. We implemented the FW clustering algorithm using C++ Boost library version 5.2, and PCA using Eigen library version 3.1.2. All experiments were performed on an Intel Core 2 Duo CPU at 2.99 GHz with 4.0 GBs memory running Ubuntu LTS (Long Term Support) OS version 12.04.

5.4 Experimental Results

In our experiments, we first collected the raw dataset for the entire run. Features were extracted and data was normalized. The data was then divided into different intervals. The first interval represented the training interval and was set to 500 s. The intervals immediately followed the first interval represented the test intervals. Each test interval was set to 1,000 s. To test the performance of the algorithms, two statistics were computed: detection rate (DR) and false alarm rate (FAR). The DR is defined as the percentage of correctly detected attacks. The FAR is defined as the percentage of incorrect instances that were classified as intrusions.

Prior studies suggested that adaptive normal profile method increased detection accuracy than a static normal profile method [2, 11]. However, there was a lack of a detailed analysis on the time interval or window size over which samples were collected. In the first set of experiments, we evaluate different time slots for 30 nodes (max speed = 5 m/s). Figure 1 depicts the performance of K-means algorithm when different time slots are applied. When detecting individual attack, we found that the time slot that yielded the best performance varied among different algorithms.

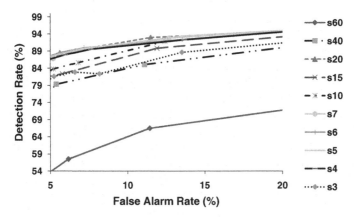

Fig. 1 Performance of K-means algorithm using different time slots

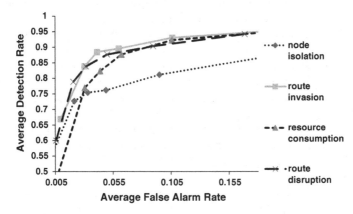

Fig. 2 FW performance comparison for different widths

On average, the results show that 5 or 6 s time slot delivers a better average performance for all techniques. From this figure, we can see that 7 and 20 s time slots also deliver good results as the lines are overlapping one another. However, FARs were slightly worse than that in 5 or 6 s time slot.

Figure 2 shows a comparison of the performance of FW for different cluster widths, which determine how close two instances have to be so that they can be assigned to the same cluster. Our results indicated that threshold values had a significant impact on the performance of the clustering techniques. Different from results reported in [11], with the same experimental parameters (e.g. 100 nodes) and time slot, when applied different thresholds, our results were better for K-means, C-means, FW, and PCA, but worse for OCSVM. For example, on detecting the flooding attacks (e.g. resource consumption attack), the DR was 83.0%, 84.7%, 78.4%, 79.8% and the corresponding FAR was 4.82, 4.82, 5.73, and 8.06% for K-means, C-means, FW, and PCA, respectively. While our DRs were comparable to that in [11], which DR and FAR were equal to 84.16 and 17.75 %, respectively, our FARs were at least 9.0 % lower. For OCSVM, although the DR was 99.2 %, the FAR remained high at 37.1 %. However by using a threshold value of 0.5, the DR and FAR decreased to 70.69 and 3.97 % respectively. The FAR is also significantly lower than that of [11].

In another set of experiments that used the same experimental parameters as those proposed in [2], our results showed that indirect application of FW provided a comparable performance reported in [8]. When applied a width of 0.2, on detecting the flooding attack, our results showed a DR of 91 % and FAR of 9.08 %, versus DR of 94.27 % and FAR if 7.82 % as reported in [2]. However, it is expected that our technique would consume less resources than those reported in [2]. The reason is that our method neither requires any retraining nor updates to the normal profile, which normally lead to additional processing time, CPU usage, and memory consumption.

Recent clustering methods proposed for anomaly detection in MANETs applied either indirect mode [2, 4, 11, 16], or direct mode [9]. It is unclear if direct mode clustering application in general provides a comparable performance to clustering technique that uses indirect mode. To test this, in another set of experiments, we evaluated the performance of each clustering technique in direct mode versus indirect mode using a simulation of 100 nodes. The results showed that, for K-means, the average DR for all 4 attacks was 81.9 and 79.2 %, and the FAR was 4.6 and 5.3 % for indirect mode and direct mode, respectively. The results were quite similar to those of C-means and FW. The average DR for all 4 attacks was 83.5 %, 84.0 %, 79.0 %, 83.7 %, and the average FAR was 4.8, 10.4, 5.5, and 10.7 % for indirect mode C-means, direct mode C-means, indirect mode FW, and direct mode FW, respectively. As can be seen, the FAR of the direct mode is almost twice as much as that of the indirect mode. This suggests that the direct mode is a preferred method for MANET intrusion detection.

In another set of experiments, the clustering techniques as well as PCA and OCSVM were compared with each other by indirect application on test datasets. The results are shown on Tables 3, 4 and 5. Compared to the other detection techniques, without tuning, OCSVM achieved the highest DR, but its FAR increased by more than 28 %. As the network expands, the performance of PCA degraded for

Table 3 Performance of each technique for N = 200

	NI		RI		RC		RD	
	DR	FAR	DR	FAR	DR	FAR	DR	FAR
K-means	82.1	5.28	88.9	5.45	87.6	5.29	87.8	5.05
C-means	82.5	5.49	89.3	5.63	88.4	5.52	88.4	5.24
FW	78.2	5.57	85.3	5.53	81.3	4.80	83.3	5.59
PCA	70.2	5.19	71.3	6.18	64.5	6.73	64.4	8.18
OCSVM	98.3	37.2	99.3	37.1	99.9	37.1	100.0	37.8

(*Notes NI* Node isolation attack; *RI* Route invasion attack; *RC* Resource consumption attack; *RD* Route disruption attack; *DR* Detection rate; *FAR* False alarm rate)

Table 4 Performance of each technique for N = 100

	NI		RI		RC		RD	
	DR	FAR	DR	FAR	DR	FAR	DR	FAR
K-means	75.2	4.81	81.3	4.38	83.0	4.82	88.3	4.59
C-means	76.8	4.90	83.0	4.58	84.7	4.82	89.4	4.80
FW	72.5	5.82	77.1	4.60	78.4	5.73	88.1	5.87
PCA	74.2	7.57	76.2	6.42	79.8	8.06	86.2	8.46
OCSVM	97.9	37.6	96.6	37.4	99.2	37.1	98.6	38.4

Table 5 Performance of each technique for N = 30, max speed = 5 m/s

	NI		RI		RC		RD	
	DR	FAR	DR	FAR	DR	FAR	DR	FAR
K-means	78.1	3.61	86.5	3.57	86.9	4.12	84.2	3.84
C-means	78.5	3.69	87.7	3.71	87.3	4.23	85.8	4.13
FW	72.7	2.14	83.8	3.02	76.9	3.05	78.8	2.05
PCA	80.0	4.36	90.5	5.67	90.4	4.34	85.0	4.64
OCSVM	99.2	34.5	99.6	34.1	98.8	34.8	98.5	32.6

at least 5 %. However, there was less variation found in the performance of other detection techniques.

In the next set of experiments, it was observed that for the same network size, all techniques performed better with lower mobility. The experiments involved 30 nodes, with different velocities (i.e., 20, 10, and 5 m/s) and pause times (i.e. 10, 50, 100, and 200 s) that simulate high, medium, and low mobility. The average DRs and FARs were then computed for each velocity. Figure 3 shows performance of all techniques declined gradually with respect to increasing velocity. For all techniques, the DRs dropped by about 10 % when the velocity increased from 5 to 20 m/s. At higher velocities, the FARs increased marginally for all techniques.

Figure 4 shows a comparison of performance of all techniques in a 30 node network with a max velocity of 20 m/s. By comparing the curves, it can seen that among all unsupervised learning techniques, K-means and C-means deliver the best results. As the DRs for K-means and C-means reached 90 %, the FARs were low compared to other techniques. Furthermore, the results of K-means, C-means, and OCSVM were also better than those reported in [2] with respect to the static normal profile results. The results also indicated that FW performed slightly worse than other unsupervised learning techniques. On average, we found that K-means required the least resource, and C-means required the most resource.

For the same simulation parameters (30 nodes, max velocity = 20 m/s), when detecting individual attacks, PCA performed best in detecting node isolation attacks (Fig. 5); OCSVM performed best in detecting route invasion attacks (Fig. 6); C-means

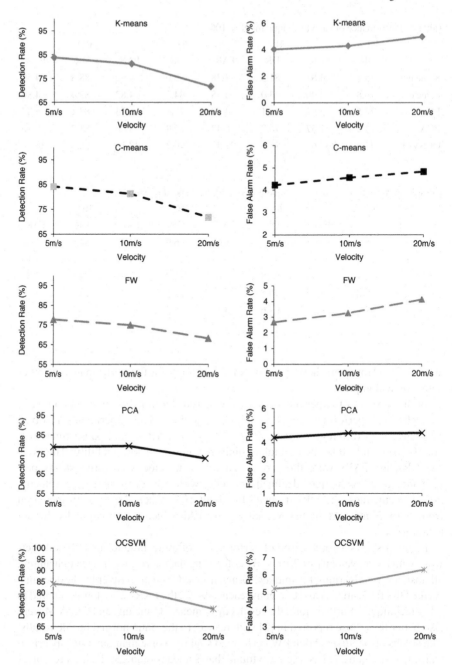

Fig. 3 Performance comparison of each algorithm using different velocities

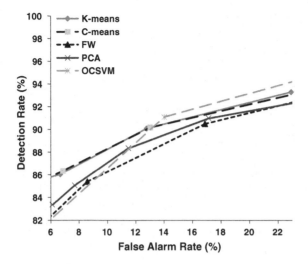

Fig. 4 Comparison of performance of all techniques in a 30 node network

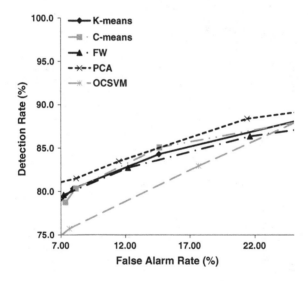

Fig. 5 Performance comparison of all techniques for node isolation attacks

performed best in detecting resource consumption attacks (Fig. 7), and K-means and C-means performed best in detecting route disruption attacks (Fig. 8). By comparing the figures, it is clear that node isolation was the hardest attack and route disruption attacks was the easiest to detect.

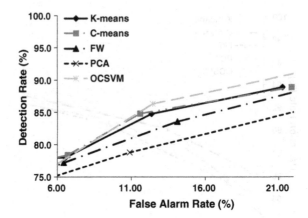

Fig. 6 Performance comparison of all techniques for route invasion attacks

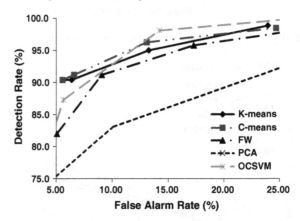

Fig. 7 Performance comparison of all techniques for resource consumption attacks

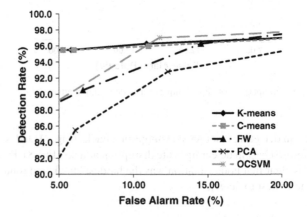

Fig. 8 Performance comparison of all techniques for resource consumption attacks

6 Conclusions

In this paper, we showed an ongoing research effort by reporting a set of experiments on unsupervised techniques with ns2 generated datasets. This is the first step towards a comprehensive empirical evaluation of unsupervised intrusion detection techniques in MANETs. Our experiments showed that a time slot of 5 s was preferred for all detection techniques in the experiments. We also found that the performance of a direct application of clustering techniques was worse than that of trained clusters. The results also indicated that K-means was the most effective and efficient detection techniques in term of performance and resource consumption, although it seems to be a rather straightforward unsupervised learning algorithm.

For future work, we plan to investigate how the mobility of the network and different normal profile selection models such as adaptive, recent, and random normal profile selection affect the performance of the detection algorithms.

References

1. Abdi, H., Williams, L.J.: Principal component analysis. Wiley Interdisc. Rev. Comput. Stat. **2**(4), 433–459 (2010)
2. Alikhany, M., Abadi, M.: A dynamic clustering-based approach for anomaly detection in AODV-based MANETs. In: Proceedings of International Symposium on Computer Networks and Distributed Systems (CNDS), pp. 67–72 (2011)
3. Anantvalee, T., Wu, J.: A survey on intrusion detection in mobile ad hoc networks. In: Xiao, Y., Shen, X., Du, S. (eds.) Wireless/Mobile Network Security, pp. 170–196 (2006)
4. Avram, T., Oh, S., Hariri, S.: Analyzing attacks in wireless ad hoc network with self-organizing maps. In: Proceedings of the 5th Annual CCNSR, pp. 166–175 (2007)
5. Azer, M. A., El-Kansa, S. M., El-Soudani, M. S.: Security in ad hoc networks from vulnerability to risk management. In: IEEE 3rd International Conference on Emerging Security Information, Systems and Technologies, pp. 203–209 (2009)
6. Chang, C., Lin, C.: LIBSVM: a library for support vector machines. ACM Trans. Intell. Syst. Technol. **2**(3), 1–27 (2011)
7. Gan, G.: Data Clustering in C++: an object-oriented approach. Chapman & Hall/CRC, Boca Raton (2011)
8. Jain, A.K., Murty, M.N., Flynn, P.J.: Data clustering: a review. ACM Comput. Surv. **31**(3), 264–323 (1999)
9. Li, W., Joshi, A.: Outlier detection in ad hoc networks using Dempster-Shafer theory. In: Proceedings of the 10th ICMDM, pp. 112–121 (2009)
10. Mitrokotsa, A., Tsagkaris, M., Douligeris, C.: Intrusion detection in mobile ad hoc networks using classification algorithms. In: Processing of IFIP International Federation for Information, Advances in Ad Hoc Networking, Vol. 265, pp. 133–144 (2008)
11. Nakayama, H., Kurosawa, S., Jamalipour, A., Nemoto, Y., Kato, N.: A dynamic anomaly detection scheme for AODV-based mobile ad hoc networks. IEEE Trans. Veh. Technol. **58**(5), 2471–2481 (2009)
12. Ning, P., Sun, K.: How to misuse AODV: a case study of insider attacks against mobile ad-hoc routing protocols. Proc. Ad Hoc Netw. **3**(6), 795–819 (2005)
13. Perkins, C., Belding-Royer, E., Das, S.: Ad hoc on-demand distance vector routing. RFC 3561 (2003)

14. Portnoy, L., Eskin, E., Stolfo, S.: Intrusion detection with unlabeled data using clustering. In: Proceedings of the ACM Workshop on Data Mining Applied to Security, pp. 1–14 (2001)
15. Sadoddin, R., Ghorbani, A.: A comparative study of unsupervised machine learning and data mining techniques for intrusion detection. Machine Learning and Data Mining in Pattern Recognition. In: Perner, P. (ed.) Vol. 4571, pp. 404–418 (2007)
16. Sun, B., Wu, K., Pooch, U.: Towards adaptive intrusion detection in mobile ad hoc networks. In: Proceedings of IEEE GLOBECOM, pp. 3551–3555 (2004)

Live Migration Performance Modelling for Virtual Machines with Resizable Memory

Cho-Chin Lin, Zong-De Jian and Shyi-Tsong Wu

Abstract The quality of services is a major concern to the users on cloud computing platforms. Migrating non-stop services across hosts on a cloud is important for many critical applications. Live migration is a useful mechanism for virtual machines to minimize service downtime. Thus, the costs of iterative pre-copy and downtime experienced by a real time application should be as short as possible. In this paper, a general model for live migration is presented. An effective strategy for optimizing the service downtime under this model is suggested. The performance of live migration is evaluated for virtual machines with resizable memory. Our results show that the service downtime can be significantly reduced by dynamically resizing memory size according to the working sets of running tasks

Keywords Live migration · Performance · Virtual machine · Resizable memory

1 Introduction

Virtualization enables system performance consolidated as well as resource utilization flexible [1]. The virtual machine monitor sits at the layer below the virtual machines and is responsible for managing resource sharing among the virtual machines. Based on the modified or non-modified kernels of guest operating systems, the virtualization mechanisms are classified into three approaches: full virtualization, para-virtualization and hardware assisted virtualization [2].

Recently, the concept of virtual machines has attracted many research interests due to its advantages in deploying the architectures for cloud computing. One of

C.-C Lin (✉) · Z.-D Jian · S.-T Wu
Department of Electronic Engineering, National Ilan University, Yilan, Taiwan
e-mail: cclin@niu.edu.tw

Z.-D Jian
e-mail: r0042002@ms.niu.edu.tw

S.-T Wu
e-mail: stwu@niu.edu.tw

© Springer International Publishing Switzerland 2015
R. Lee (ed.), *Computer and Information Science*, Studies in Computational
Intelligence 566, DOI 10.1007/978-3-319-10509-3_7

the advantages is that it guarantees non-stop services by performing virtual machine migration as it is necessary. Live migration iteratively duplicates the page frames of a virtual machine across hosts. Since intensive page frame duplication can adversely impact the performance of a system, an appropriate strategy is necessary to minimize service downtime. In this paper, a general model for live migration is presented. Under this model, we suggest an effective strategy for minimizing the downtime of live migration. We also conduct experiments with various parameter settings to show the usefulness of our approach. Our results show that downtime can be significantly reduced by dynamically resizing memory size according to the working sets of running tasks.

The paper is organized as follows. The related works are discussed in Sect. 2. In Sect. 3, the general model of live migration is defined. In Sect. 4, the adoptability of our model to the well known Xen hypervisor is discussed. In Sect. 5, an effective strategy inspired by the experiments is suggested. Finally, concluding remarks and future research directions are provided in Sect. 6.

2 Related Works

A technique for migrating in-service OS instances using bounded iterative push phase was developed to achieve acceptable total migration time and downtime on fast networks [3]. In [4], post-copy approach which duplicates memory pages on faults was proposed. Although the downtime of post-copy was small, page faults degrade the performance of virtual machines in general. In addition, dynamic self-ballooning (DSB) was also employed to reduce the number of free pages such that the total migration time can be further reduced. The advances in live migration techniques and several open issues were provided and discussed in [5]. In [6], the authors studied the time needed by a virtual machine live migration regarding the memory size, working set and dirty rate. They showed that the time interval of stop-and-copy stage for a live migration is constant if the dirty rate exceeds a threshold value. It was observed the length of the interval depends on the size of working set during live migration. In [7], a second chance strategy was proposed for iterative pre-copy stage to reduce the total number of pages to be duplicated across hosts. The strategy can duplicate less pages in each iteration without increasing the total number of iterations needed by Xen. In [8], a technique was proposed to avoid the transmission of frequently updated pages. It was achieved by adding an additional bitmap to mark the frequently updated pages and those pages can only be duplicated in the last iteration. In [9], a technique of identifying unused pages and encoding duplicated pages was used to reduce the size of pages to be duplicated across hosts during live migration.

3 Model of Live Migration

In general, live migration of a virtual machine needs to duplicate the content of memory frames across hosts while the virtual machine is performing computations. It is obvious that a live migration proceeds with two overlapped activities: page-accessing activity and page-duplicating activity. The page-duplication activity proceeds according to a predefined protocol until a terminating condition has been satisfied. Thus, the model of live migration for virtual machine $\mathcal{V}$ can be defined as $\mathcal{V}(\mathcal{M}, P, C)$, where $\mathcal{M}$ is memory configuration, P is duplication protocol and C is a set of terminating conditions. Memory configuration describes the sizes of various memory spaces which can affect the performance of a live migration. It is a triple $\mathcal{M} = (M_{max}, M_{avl}, M_{tsk})$, where M_{max} is the maximum memory size of virtual machine $\mathcal{V}$, M_{avl} is the memory size available to the guest OS and running tasks, and M_{tsk} is the memory size needed by the guest OS and running tasks. If $M_{avl} < M_{tsk}$, then auxiliary storage is needed to provide an extra space such that the task can start successfully. During the live migration, each page frame of the virtual machine is checked periodically to identify whether it has been modified most recently. The successive checks form a sequence which identifies that page frame modification history. Let h_k indicate the modification status of a frame. Then, $h_k = 1$ if the page frame is modified at iteration k; otherwise, $h_k = 0$. The sequence $h_1 h_2 h_3 \ldots h_k$ forms the pattern which indicates the history of updating the frame at the interval of the first iteration and the kth iteration. History observation window $\hbar$ defines the number of most recent iteration checked for the need of page duplication during a live migration. For example, if $\hbar = 3$ then the subsequence $h_{k-2} h_{k-1} h_k$ is extracted from history pattern at the end of kth iteration and it is compared with the predefined patterns for page duplication. The predefined patterns are defined in the duplication protocol as $P = (\hbar, \mathcal{D})$, where set $\mathcal{D}$ defines the patterns used to trigger a page duplication activity. For example, if $P = (3, \{010, 001\})$ then the frames with $h_{k-2} h_{k-1} h_k = 010$ or 001 will be duplicated. Notice that the protocol only applies to the page frames of available memory. C is a set of rules for terminating iterative page duplication and starting crossing-host memory consistency.

4 Live Migration on Xen

In this section, we will show that the live migration of a virtual machine under Xen can be defined using our model. The first parameter of live migration model is memory configuration $\mathcal{M}$ which consists of three elements M_{max}, M_{avl} and M_{tsk} as illustrated in Fig. 1. Figure 1a shows that the whole memory is divided into three overlapped areas: maximum memory M_{max}, available memory M_{avl} and task memory M_{tsk}. The sizes of memory areas M_{max}, M_{avl} and M_{tsk} are 38 frames, 30 frames and 21 frames, respectively. In the figure, eight memory frames marked with gray color are the area blocked by balloon driver. Even though 38 memory frames are given to the virtual

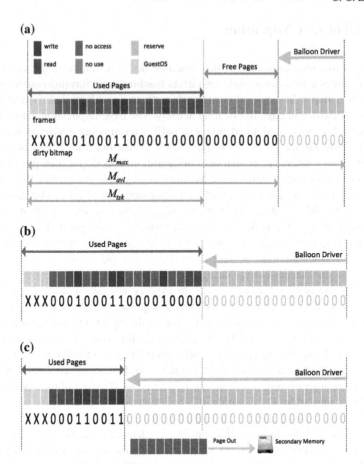

Fig. 1 Memory configuration of a virtual machine with resizable memory. **a** $M_{avl} > M_{tsk}$. **b** $M_{avl} = M_{tsk}$. **c** $M_{avl} < M_{tsk}$

machine when it is created, the blocked area is not available to guest OS and any running task. Thus, no page of tasks can be allocated to the frames in the blocked area. It implies that the blocked area does not involve in the page-duplication activity while a virtual machine is migrating across hosts. Nine memory frames marked with blue color are the free area which is not used by guest OS and any running task currently. Any memory frame in the free area is still available to a running task whenever more memory space is needed by the task. Thus, the free area should involve in the page-duplication activity. Since the memory is resizable, the available memory M_{avl} can be adjusted using balloon driver as illustrated in Fig. 1b. In the figure, the available memory M_{avl} has been reduced to 21 frames which can just accommodate the guest OS and the running task. The available memory can be further reduced to 12 frames as shown in Fig. 1c by enlarging the blocked area to 26 frames using balloon driver. It is obvious that the number of frames is not large enough to accommodate both the guest OS and the running task. In this case, a secondary storage is used to provide an extra space to accommodate the running task.

Page duplication protocol is developed to keep memory as consistent as possible while a virtual machine is migrating across hosts. The second parameter of live migration model is duplication protocol P which consists of two elements $\hbar$ and $\mathcal{D}$. Pre-copy strategy is employed by Xen to migrate a virtual machine across hosts with non-stop service. While a virtual machine is in an iterative pre-copy stage, two activities proceed concurrently: page-accessing activity and page-duplicating activity. In each iteration, page-accessing activity at the source host may access the pages in the frames which may make the pages to be duplicated across hosts. Xen uses three bitmaps to_skip, to_send and to_fix to define the duplication protocol [10]. Bitmaps to_skip[j]=1 and to_send[j]=1 indicate the jth frame has been updated in this and the previous iterations, respectively. The necessity of duplicating page P_j is determined by the j entries of the bitmaps to_skip and to_send. If to_skip[j]=0 and to_send[j]=1 then page P_j will be sent. The flowchart for Xen's duplication protocol is illustrated in Fig. 2. Each iteration needs to scan bitmaps to_skip and to_send once. In the iteration, the pages to be duplicated across hosts are collected to a buffer and sent to target host in several rounds. The second parameter of live migration under Xen can be represented by $P = (2, \{10\})$.

The third parameter of live migration model is a set of terminating conditions C. Four terminating conditions are given by Xen to avoid an endless pre-copy process [11]. The iterative pre-copy stage terminates and the stop-and-copy stage starts if and only if one of the conditions is satisfied. Condition C1: the number of updated frame in an iteration is less than 50; condition C2: the number of iterations has reached 29; condition C3: the number of total duplicated pages exceeds the threefold of the maximum memory space M_{max}. Condition C4: the number of duplicated pages in this iteration is larger than that in the previous iteration and the measured network bandwidth reaches its maximal value. Condition C1 guarantees a short downtime since only a few pages are to be duplicated in the stop-and-copy stage. Conditions C2 and C3 force the iterative pre-copy stage to terminate and go to the stop-and-copy stage in case of condition C1 never occurs. In general, Condition C4 has default value equal to false. Thus, the set of terminating conditions C ={C1, C2, C3}.

Two of the performance meters for virtual machine live migration are the total migration time and downtime [5]. The total migration time T_{mgt} can be expressed using the following equation [11]:

$$T_{mgt} = T_a + \sum_{i=1}^{n} t_{copy}(i) + T_b \tag{1}$$

T_a and T_b are the overheads ahead and behind the stage of iterative pre-copy, respectively. T_a includes the time for initialization and resource reservation. T_b includes the time for commitment and virtual machine activation. In the equation, $t_{copy}(i)$ is the time needed by duplicating pages in the ith iteration. Thus, $\sum_{i=1}^{n-1} t_{copy}(i)$ is the total pre-copy time if there are n iterations in the live migration. Note that the nth

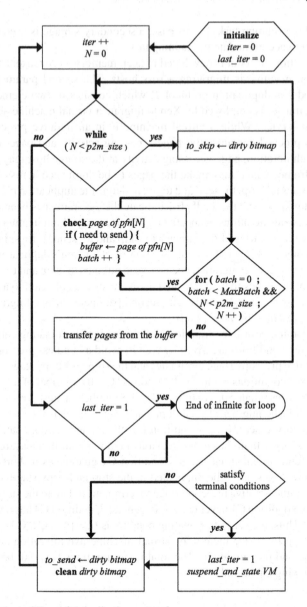

Fig. 2 Flowchart of Xen 4.0.1 duplication protocol

iteration is the final iteration in which the virtual machine at the source host stops providing services. The sum of T_b and $t_{copy}(n)$ is also referred as service downtime. The time T_b is independent of the page-accessing pattern and can be considered as a constant. However, the time $t_{copy}(n)$ which can be experienced by the service requesters is dependent on the number of pages to be duplicated in the last iteration.

Thus, an effective strategy which leads to a favorable condition is needed. In this paper, we will show how the memory space affects the total migration time T_{mgt} and the final page-duplication time $t_{copy}(n)$. Let M be the size of memory space given to a virtual machine and m be the memory frame size. In [11], the total migration time and service downtime can be bounded using the following expression.

$$\frac{M}{B} + T_o \leq T_{mgt} \leq \frac{5M - m}{B} + T_o \qquad (2)$$

$$T_b \leq T_{dt} \leq \frac{M}{B} + T_b \qquad (3)$$

where $T_o = T_a + T_b$, B is the network bandwidth and T_{dt} is the downtime. Note that $(T_{mgt} - T_o)$ is the time needed by page duplication during live migration. Thus, $(T_{mgt} - T_o) \times B$ is the number of total bytes to be duplicated across hosts. From Eq. (2), we have that the lower bound and the upper bound on the numbers of duplicated bytes are M and $5M - m$, respectively. $(T_{dt} - T_o)$ is the time for the final round of page duplication. From Eq. (3), we have that the lower bound and the upper bound on the times of final page duplication activity are 0 and $\frac{M}{B}$. The DSB proposed in [4] and adopted in Xen has been used to adjusted the available size of the allocated memory for achieving efficient migration. The bounds are important to develop an efficient technique for reducing the total migration time and service downtime. Although the expressions given by Eq. (2) and Eq. (3) seem to have intuitively captured the cost of the live migration for a virtual machine, more precise bounds should be found by classifying the memory space into various groups according to the usages. In the next section, we will refine the expressions based on our model.

5 Performance Measurement and Discussion

In this section, we modify Eqs. (2) and (3) in Sect. 4 using the first parameter of our proposed model. The performance of live migration is studied by varying the sizes of memory areas and the patterns of page-accessing.

5.1 Bounds on Live Migration

The time bounds on T_{mgt} and T_{dt} are important indicators for measuring the efficiency of migrating a virtual machine across hosts. Thus, it is important to derive a precise time bounds on T_{mgt} and T_{dt} before we proceed to develop an efficient strategy of live migration. To have precise bounds on T_{mgt} and T_{dt}, several experiments are conducted to capture the courses of live migrations under various parameter settings. In our experiments, the source and target hosts are networked by 100 Mbps Ethernet and the physical memory of each host is 4096 MB. NFS server is installed at the source host such that the image files of virtual machines are accessible by any remote host. The virtual machines have Xen 4.0.1 as their virtual machine monitors and the memory of each virtual machine is 2048 MB when it is created, that is, $M_{max} = 2048$ MB. The guest OS of each virtual machine occupies about 110 MB. The size of each task of various types to be executed in a virtual machine is tailored to be about 890 MB. Thus, the second parameter of our model M_{tsk} is 1024 MB. In our experiment, the virtual machine which accommodates a running task and the guest OS migrates across the hosts and the courses of the live migrations are investigated.

The first type of task performs *memory-write* periodically and two consecutive *writes* are performing in two consecutive memory frames. Note that the size of a memory frame is 4 KB. The courses of live migrations are analyzed under various frame dirty rates. The dirty rate is defined as $\frac{d(t)}{t}$, where $d(t)$ is the number of writes during the time interval t. The $d(t)$ is measured under a large available memory area which can accommodate the guest OS and the running task. Thus, the task can perform work without any page-swapping activity. Note that the actual dirty rate is lower if $M_{avl} < M_{tsk}$ due to the adverse effect of page-swapping. We implement dirty rate $d(t)$ by inserting a finite loop between write operations. In our experiment, balloon driver has been employed to resize the available memory area to $M_{avl} = 512, 1024$ or 2048 MB. We denote V_{512}, V_{1024} and V_{2048} as the virtual machines with $M_{avl} = 512$, 1024 and 2048 MB, respectively.

The total migration times for the different sizes of available memory areas using various dirty rates are given in Fig. 3. Note that for $M_{avl} = 512$ MB, the page-swapping activity occurs and the actual dirty rate is less than the dirty rate given on the x axis. Each value is plotted by averaging five time experiments on the same parameter setting. From the figure, it is observed that the minimum time of live migration for V_{512} is less than those of V_{1024} and V_{2048}. As expected, the maximum time of live migration for V_{512} is also less than those of V_{1024} and V_{2048}. The virtual machine of $M_{avl} = 512$ MB constantly swaps pages between memory and secondary

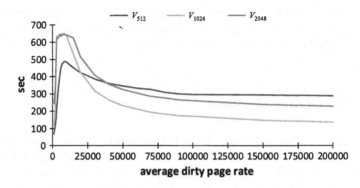

Fig. 3 Total migration time

storage due to the shortage of available memory area. This leads to the situation which satisfies terminating condition C1 for a very low dirty rate. Each of the virtual machines V_{1024} and V_{2048} sends a large amount data in an iteration if the dirty rate is close to the network bandwidth. Thus, the live migration terminates due to the number of duplicated pages exceeding threefold the maximal memory space M_{max}. When the dirty rate is low, the total migration times of V_{1024} and V_{2048} dominate V_{512}. The reason is that the shortage of memory space limits the number of pages to be duplicated between hosts. However, when the dirty rate is high, the constantly page-swapping decreases the actual dirty page rate. According to the rule set by Xen monitor, a dirty frame will be skipped without being duplicated if it is updated in the most recent iteration. Thus, V_{512} spends more time for live migration for a high dirty rate. V_{2048} dominates V_{1024} since all the frames in the free memory area are sent in the first iteration.

The total iteration numbers of the live migrations for V_{512}, V_{1024} and V_{2048} are illustrated in Fig. 4. In the figure, the vertical line which crosses the x-axis partitions the figure to a left region and a right region. The left area indicates the virtual machines work with a dirty rate lower than the network bandwidth measured in pages per second. Similarly, the right area indicates the virtual machines work with a dirty rate higher than the network bandwidth. In the left region, it is observed that the number of iterations increases in proportion to the dirty rate for V_{512}, V_{1024} and V_{2048}. Consider the case of dirty rate equal to network bandwidth: virtual machine V_{512} stops iterative pre-copy when the iteration number reaches thirty. Contrary to virtual machine V_{512}, the iteration numbers taken by V_{1024} and V_{2048} are no more than thirteen. It implies that the number of duplicated pages exceeds threefold of maximum memory space M_{max} before condition C2 is satisfied. Thus, we conclude that a virtual machine which performs work with a large memory area under just enough network bandwidth takes a long time to migrate. The reason is that a page is always checked before it is updated again. Thus, history pattern "10" is always detected in each iteration. Note that the states of all the pages will be reset to clean

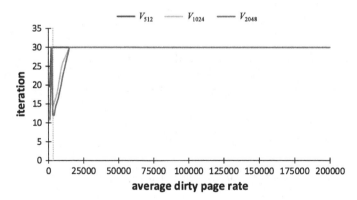

Fig. 4 Total iteration number

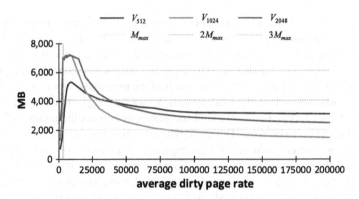

Fig. 5 Total duplication amount

before the beginning of each iteration as illustrated in Fig. 2. However, V_{512} steadily increases the iteration number due to the effect of page-swapping. In the right region, the iteration numbers of V_{1024} and V_{2048} are thirty. The reason is that a high dirty rate makes it highly possible to update a page twice before it is checked.

The total numbers of bytes to be duplicated between hosts are given in Fig. 5. From the figure, it is observed that the maximum number of duplicated bytes for the virtual machines is no more than $3.5M_{max}$. The number is bounded by $5M_{max} - m$ which satisfies the expression given in Eq. (2), if M means the maximum memory space M_{max}. Note that m is the size of a memory frame. The reason that the gap between our number and the upper bound is the task does not occupy all the memory space M_{max}. However, the minimum number of duplicated bytes is not bounded by M_{max} for V_{512} under a very low dirty rate as shown in Fig. 5. Thus, a modification to the lower bound given by Eq. (2) is necessary. It is obvious that the memory pages classified as M_{avl} should be sent at least once. We define the M in the lower bound term as the available memory area M_{avl}. The refined expression is given as follows:

$$\frac{M_{avl}}{B} + T_o \leq T_{mgt} \leq \frac{5M_{max} - m}{B} + T_o \qquad (4)$$

Furthermore, terminating condition C1 will be satisfied soon if the dirty rate is very lower. Figure 6 illustrates that the number of bytes to be duplicated in the final iteration depends on the minimum of M_{avl} and M_{tsk}. Thus, the refined expression is given as follows:

$$T_b \leq T_{dt} \leq \frac{\min\{M_{avl}, M_{tsk}\}}{B} + T_b \qquad (5)$$

It implies that the number of total duplicated bytes also depends on the minimum of M_{avl} and M_{tsk}.

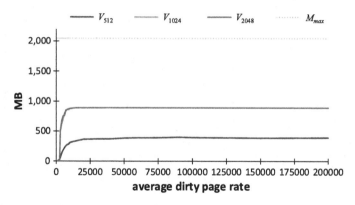

Fig. 6 Duplication amount at the last iteration

5.2 Live Migration Under Reading Access

In this section, we study the performance of live migration for a task without performing memory-updating under various sizes of available memory areas. The second type
of task only performs *memory-read* periodically and two consecutive *reads* are performing in two consecutive memory frames. The consecutive readings are performed
in the round-robin style. The number of total bytes to be duplicated across hosts are
given in Fig. 7. The figure shows that the number of bytes to be duplicated across
hosts depends on the size of available memory area M_{avl}. In the figure, the minimum
total migration time can be observed for $M_{avl} = M_{tsk}$. Note that $M_{tsk} = 1024$ MB.
If a page is swapped into a frame from the secondary storage, the frame is considered as a dirty frame and it will be duplicated across hosts in later time. When
$M_{avl} < M_{tsk}$, the total migration time increases in proportion to the size of available
memory. The reason is that if the available memory area becomes larger, more dirty
frames can remain in the memory space and will be duplicated across in later time.

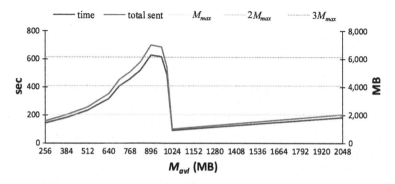

Fig. 7 Total migration time and total duplication amount under readings

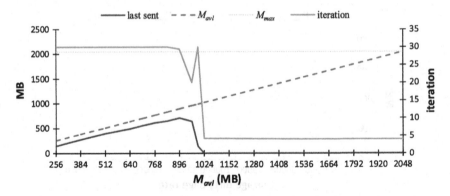

Fig. 8 Total iteration and duplication amount at the last iteration under readings

When $M_{avl} \geq M_{tsk}$, it is observed that the total migration time is also in proportion to the size of available memory in a much slower rate compared with the case of $M_{avl} < M_{tsk}$. In the case of $M_{avl} > M_{tsk}$, all the referenced pages stays in the memory space and the pages in the available memory area are sent in the first iteration. The gap of the total migration time is due to the difference in the number of free pages which are sent in the first iteration.

The number of iterations and the number of bytes to be duplicated across hosts in the last iterations is given in Fig. 8. From the figure, it is observed that the number of iterations keeps thirty for $M_{avl} < M_{tsk}$. However, the number of bytes to be duplicated across hosts increases in proportion to the size of available memory for the same case. The reason is that the size of available memory area is small, thus, the number of bytes to be duplicated is limited. It causes that the final iteration is triggered by satisfying terminating condition C2. For $M_{avl} \geq M_{tsk}$, the number of iterations is close to five. The reason is that there are no dirty pages and no page-swapping events. Thus, there are almost no pages to be duplicated across hosts in the last iteration and the situation which leads to the final iteration is satisfied by terminating condition C1.

5.3 Memory Resizing

The memory space is classified into three areas as shown in Fig. 1. In Xen, the pages in the memory frames which are blocked by balloon driver will not be duplicated across hosts in any iteration. Available memory area which is not large enough to accommodate the guest OS and running task may cause page-swapping activity. The page stored in NFS system is not duplicated to the target host. So, it is possible to reduce the number of pages to be duplicated across hosts by taking the advantage of NFS systems. The free page marked with 0 as illustrated in Fig. 1a should be sent in the first iteration. If we shrink the available memory area, the number of page

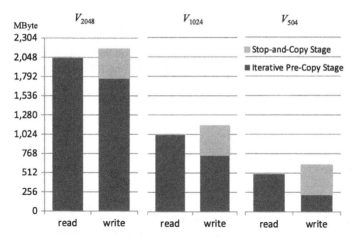

Fig. 9 Memory resizing effect for reading/writing

to be duplicated across hosts can be reduced as shown in Fig. 1b. It implies that an effective strategy of live migration can be achieved by pushing out all the no access page by shrinking the available memory. Thus, the number of page to be duplicated across hosts can be minimized

To show the correctness of our statements, we conduct experiments using various memory size settings. In the experiments, two tasks of different memory accesses are provided. Each of the tasks needs 890 MB memory space and only access to a area of 390 MB in the 890 MB memory space. That is, the working sets of the tasks are 390 MB. The first task only writes data to memory and the second task only reads data from memory. Note that the size of guest OS is about 100 MB. Thus, M_{tsk} is about 1000 MB. We claim the $M_{max} = 2048$ MB. The experimental results are given in Fig. 9. We can see that if the whole memory space is available to the task, the number of pages to be duplicated is the largest. However, if we reduce the available memory to M_{tsk}, the time can be reduced. Furthermore, if we can push out the no access data by shrinking the available memory area to 504 MB, the number of pages to be duplicated should be minimized. Note that the numbers of the duplicated pages for the three cases are the same in the final iteration since the sizes of the working sets are fixed.

6 Conclusion

In this paper, we have presented a performance model of live migration for virtual machines. Under this model, no-access-page-out strategy for reducing the costs of live migration is investigated. Experiments have been conducted using various parameter settings. It has been shown that the time of live migration can be reduced significantly if the memory space is resized dynamically based on the set of no-access pages. In the future, an algorithm will be developed for predicting the number of duplicated pages in the final iteration. Based on the prediction, the available memory area can be adjusted to minimize the downtime for real time applications.

Acknowledgments This research is supported by National Science Council under the grant 102-2221-E-197-011.

References

1. Barham, P., Dragovic, B., Fraser, K.: Xen and The Art of Virtualization. In: Proceedings the 19th ACM Symposium Operating Systems Principles, pp. 164–177 (2003)
2. The VMware Team: Understanding Full Virtualization, Paravirtualization, and Hardware Assist. http://www.vmware.com/files/pdf/VMware_paravirtualization.pdf (cited April 2014)
3. Clark, C., Fraser, K., Hand, S.: Live Migration of Virtual Machines. In: Proceedings the 2nd Symposium Networked Systems Design and Implementation, pp. 273–286 (2005)
4. Hines, M. R., Gopalan, K.: Post-Copy Based Live Virtual Machine Migration Using Adaptive Pre-Paging and Dynamic Self-Ballooning. In: Proceedings the ACM SIGPLAN/SIGOPS Int'l Conference Virtual Execution Environments, pp. 51–60 (2009)
5. Strunk, A.: Cost of Virtual Machine Live Migration: A Survey. In: Proceedings IEEE the 8th World Congress on Services, pp. 323–329 (2012)
6. Salfner, F., Tröger, P., Polze, A.: Downtime Analysis of Virtual Machine Live Migration. In: Proceedings the 4th Int'l Conference Dependability, pp. 100–105 (2011)
7. Lin, C.-C., Huang, Y.-C., Jian, Z.-D.: A Two-phase Iterative Pre-copy Strategy for Live Migration of Virtual Machines. In: Proceedings the 8th Int'l Conference Computing Technology and Information Management, pp. 29–34 (2012)
8. Ma, F., Liu, F., Liu, Z.: Live Virtual Machine Migration Based on Improved Pre-copy Approach. In: Proceedings IEEE Int'l Conference Software Engineering and Service Sciences, pp. 230–233 (2010)
9. Ma, Y., Wang, H., Dong, J.: ME2: Efficient Live Migration of Virtual Machine With Memory Exploration and Encoding. In: Proceedings IEEE Int'l Conference Cluster Computing, pp. 610–613 (2012)
10. Liu, Z., Qu, W., Yan, T.: Hierarchical Copy Algorithm for Xen Live Migration. In: Proceedings Int'l Conference Cyber-Enabled Distributed Computing and Knowledge Discovery, pp. 361–364 (2010)
11. Akoush, S., Sohan, R., Rice, A.: Predicting the Performance of Virtual Machine Migration. In: Proceedings IEEE Int'l Symposium Modeling, Analysis and Simulation of Computer and Telecommunication Systems, pp. 37–46 (2010)

A Heuristic Algorithm for Workflow-Based Job Scheduling in Decentralized Distributed Systems with Heterogeneous Resources

Nasi Tantitharanukul, Juggapong Natwichai and Pruet Boonma

Abstract Decentralized distributed systems, such as grids, clouds or networks of sensors, have been widely investigated recently. An important nature of such systems is the heterogeneity of their resources; in order to archive the availability, scalability and flexibility. As a consequence, managing the systems to meet requirements is obviously a nontrivial work. The issue is even more challenging in term of job scheduling when the task dependency within each job exists. In this paper, we address such problem of job scheduling, so called workflow-based job scheduling, in the decentralized distributed systems with heterogeneous resources. As such problem is proven to be an NP-complete problem, an efficient heuristic algorithm to address this problem is proposed. The algorithm is based on an observation that the heterogeneity of the resources can affect the execution time of the scheduling. We compare the effectiveness and efficiency of the proposed algorithm with a baseline algorithm. The result shows that our algorithm is highly effective and efficient for the scheduling problem in the decentralized distributed system with heterogeneous resources environment both in terms of the solution quality and the execution time respectively.

Keywords Decentralized algorithm · Heterogeneous resources · Workflow-based job scheduling

1 Introduction

In recent years, distributed systems such as P2P and computational grid become a generic platform for high performance computing [1]. In term of management, a distributed system can be centralized or decentralized [5]. In the centralized

N. Tantitharanukul · J. Natwichai (✉) · P. Boonma
Data Engineering and Network Technology Laboratory, Department of Computer Engineering, Faculty of Engineering, Chiang Mai University, Chiang Mai 50200, Thailand
e-mail: juggapong@eng.cmu.ac.th

N. Tantitharanukul
e-mail: n.tantitharanukul@gmail.com

P. Boonma
e-mail: pruet@eng.cmu.ac.th

© Springer International Publishing Switzerland 2015
R. Lee (ed.), *Computer and Information Science*, Studies in Computational
Intelligence 566, DOI 10.1007/978-3-319-10509-3_8

distributed system (CDS) setting, a central controller is required for coordinating resource sharing and computing activities among computing machines. That is, all the system-wide decision makings are coordinated by the central controller. Contrary to the CDSs, in decentralized distributed systems (DDSs), multiple controllers can coexist and cooperate. Thus, management services such as job scheduling, resource discovery and resource allocation can be managed by the multiple controllers. If a controller fails, the other controllers can take over its responsibility autonomously. This is the major advantage of the DDSs. Examples of the DDSs are grid and cloud computing systems.

Although DDSs have the aforementioned advantage, completely decentralized nature of DDSs raises a big challenge in job scheduling [5, 6]. In addition, for composite jobs, i.e. jobs with multiple sub-processs or tasks, the tasks in the jobs can have dependency, e.g. a task requires an output of another task as its input. Thus, some tasks are prohibited to be executed concurrently. For instance, in order to calculate the histogram of collected sensing data, data grouping need to be performed first; thus, histogram calculation depends on data grouping. Therefore, the jobs must be executed under valid flow-constraints, generally, described as a workflow template. We can represent a workflow template using a directed acrylic graph (DAG) where the tasks are represented by nodes while the dependency (the execution order) between tasks is represented by edges. Figure 1 shows examples of workflow template where t_i represents a task with index i and w_j represents a workflow template with index j.

From Fig. 1, to finish a job with workflow template w_1, the tasks must be executed in the order of t_1, t_2, and t_3. In workflow template w_2, the tasks t_5 and t_6 can be executed in parallel after t_4 is completed. In workflow template w_3, each task has no dependency on the others, in this case, the set of edges is empty.

There are several attempts on decentralized job scheduling, e.g. [2–4, 8]. For example, in [8], a decentralized scheduling algorithm for web services using linear workflow is proposed. In this work, linear workflow allows only one incoming edge for each node. The goal of this work is to minimize the response time. However, this work only focuses on linear workflow which might not be practical in the real world.

In [7], the authors proposed a workflow-based composite job scheduling for decentralized distributed systems algorithm. The algorithm is based on an observation that

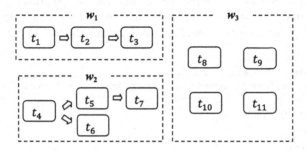

Fig. 1 Examples of workflow templates

the degree of each task significantly affects the execution time of all jobs. The result shows that this algorithm is efficient, the solution of this algorithm is mostly close to the optimal solution. However, this algorithm is evaluated only in the homogeneous resources decentralized distributed systems.

In this paper, we propose an algorithm for the mentioned problem with heterogeneous resources condition, i.e. each resource can only be used to execute some specific tasks from any workflow template. Such difference in resources often exists in the real-world scenarios e.g. the cost of some data centers might be too high for executing some tasks, or the result of some tasks might be too large to be propagated from one site to another economically. The main aim of the proposed algorithm is to minimize the total execution time when a DDS has to process multiple jobs simultaneously with such heterogeneous resources. As the problem has been proven as an NP-complete problem we propose a heuristic algorithm to find the solution.

The idea of the proposed algorithm is to allocate the resource to a task based on the heterogeneity of such resource. A resource with less heterogeneity, or more restriction, will be allocated to the task earlier. This can preserve the more heterogenous resources for the later tasks. The experimental results, which compare the proposed algorithm with a baseline algorithm in [7], is shown to illustrate the effectiveness and the efficiency of our work.

The rest of this paper is organized as follows: Sect. 2 introduces the basic definitions and formulates the **minimum length of time-slot with heterogeneous resource** **(MLTH)** problem. Section 3 presents a heuristic algorithm to find the minimum execution time. Experimental results are presented in Sects. 4 and 5 concludes this paper.

2 Basic Definitions and MLTH Problem

In this section, we introduce the basic notations and concepts based on the previous work in [7]. Then, the MLTH problem is defined with our additional heterogeneity notations.

Definition 1 (*Distributed System*) A distributed system D is presented by an undirected graph where each node corresponds to a machine in the system. The finite set $N(D)$ denotes the set of nodes in D, and the finite set $E(D)$ is the set of edges where each edge corresponds to a non-directed connection between two nodes.

Definition 2 (*Resource*) Let n_i be a node in $N(D)$, $n_i \in N(D)$. The resources of n_i are the computing units, that n_i can use to execute a computing process. The set of the resources of n_i is denoted as $R(n_i)$ whereas the set of resources of D is denated as Ω, $\Omega = \bigcup R(n_i)$.

Then, the decentralized distributed systems (DDSs) as in [5] can be defined as follows,

Definition 3 (*DDS*) Let D be a distributed system, D is classified as a decentralized distributed system (DDS) if it has multiple controllers, where the controller is a node that can allocate the resources of itself and some other nodes.

As the dependencies between some tasks processed by a DDS exist, then, tasks, workflow templates, jobs are defined as follows,

Definition 4 (*Task*) Let D be a DDS, a task in D is a unit of computing process that a node in D can complete execution in a unit of time. A set of all tasks that can be executed by the resources in Ω denotes by T.

Definition 5 (*Workflow Template*) Workflow templates in D are the directed acyclic graph where each node corresponds to a task, t_i, and each edge indicates the data flow between two tasks. Given a workflow template w_x in the set of all workflow templates W, $\overline{N}(w_x)$ denotes the node set of w_x where $\overline{N}(w_x) \subseteq T$. On the other hand, $\overline{E}(w_x)$ denotes the directed edge set in the workflow template. For any workflow template w_x, task t_l is called a predecessor of task $t_{l'}$, if and only if, the order pair $(t_l, t_{l'}) \in \overline{E}(w_x)$. This indicates that task t_l must be executed and completed before execution of task $t_{l'}$. Meanwhile, task $t_{l'}$ is called a successor of task t_l. The node without incoming edge from other nodes is called the start task. On the other hand, the node without outgoing edge to other nodes is called the end task.

From Definition 5, we use the workflow template to be the template of any job that the system can execute. It can show the execution flow of a job, So, we define the definition of a job in Definition 6.

Definition 6 (*Job*) Let W be the set of workflow templates, a job j_k in D is an instance of a workflow template w_k in W. The task t_l of job j_k is denoted by t_l^k and $\overline{T}$ is the set of all tasks from J, where J is the set of all jobs in D.

As the DDSs may have the heterogeneous resources as mentioned before, each resource can only be used to execute the specific tasks from a workflow due to some limitation, e.g. cost of the communication. For example, assume that we have 3 resources in the system denoted as r_1, r_2, and r_3 as shown in Fig. 2. In the figure, an example of the executable tasks of each resource is shown. It can be seen that, resource r_1 can be used to execute only 7 tasks which are the tasks from workflow w_1 and w_2 in Fig. 1. Meanwhile resource r_2 can be used to execute all the tasks of the jobs that are an instance of any workflow template, i.e. w_1, w_2 or w_3.

In order to allow the heterogeneity in the DDSs, we define the executable set as follows.

Fig. 2 The executable tasks of r_1, r_2, and r_3

Resource	Executable Tasks, $\lambda(r_i)$	$\|\lambda(r_i)\|$
r_1	$t_1, t_2, t_3, t_8, t_9, t_{10}, t_{11}$	7
r_2	All tasks	11
r_3	$t_4, t_5, t_6, t_7, t_8, t_9, t_{10}, t_{11}$	8

Definition 7 (*Executable Set*) Let r_p be the resource, $r_p \in \Omega$, an executable set of resource r_p is the set of tasks from all workflow templates that r_p can execute, denotes by $\lambda(r_p)$ where $\lambda(r_p) \subseteq \bigcup \overline{N}(w_x)$ where $w_x \in W$.

After the basic definitions related to the jobs have been defined, the time-slot is defined for describing the job scheduling, or the execution flow, as follows.

Definition 8 (*Time-slot*) Let J be a set of current jobs in D, the time-slot of J on Ω is the function $S : I^+ \times \Omega \to \overline{T} \cup \{null\}$ where $S(\alpha_q, r_p) \in \lambda(r_p), r_p \in \Omega$, and α_q is a time unit where $\alpha_q \in I^+$ and I^+ is the set of natural numbers.

The domain of S is the order pair of time unit α_q, $\alpha_q \in I^+$, and resource r_p, $r_p \in \Omega$. Thus the range of S, $S(\alpha_q, r_p)$, is the executed task that uses resource r_p at time unit α_q, and $S(\alpha_q, r_p) \in \lambda(r_p)$. When there is no task to be executed on resource r_p at time unit α_q, $S(\alpha_q, r_p)$ is *null*.

For any $S(\alpha_q, r_p)$ and $S(\alpha_{q'}, r_{p'})$, where $p \neq p'$, if $q = q'$ then $S(\alpha_q, r_p)$ and $S(\alpha_{q'}, r_{p'})$ are executed in parallel. If $q < q'$ then $S(\alpha_q, r_p)$ is executed before $S(\alpha_{q'}, r_{p'})$, also, $S(\alpha_q, r_p)$ is executed before $S(\alpha_{q'}, r_p)$.

In order to illustrate the concepts clearly, the time-slot can be represented as a 2-dimensional matrix. Figure 3 shows a time-slot structure, a cell at row q and column p in the structure represents $S(\alpha_q, r_p)$. We can see that the representation can illustrate the execution flow of multiple tasks from multiple jobs along with their workflow template.

In order to define the problem precisely, we also introduce the length of the time-slot, as follows.

Definition 9 (*Length of Time-slot*) The length of time-slot S is the maximum value of time unit α_q which is $S(\alpha_q, r_p)$ is not *null*, $\exists r_p \in \Omega$.

For a DDS, there can be many versions of the time-slots for a given $\overline{T}$. For example, let us reconsider the given workflow templates in Fig. 1. Suppose that there are five

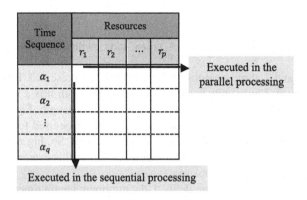

Fig. 3 The structure of timeslots

Fig. 4 Workflow template of j_1 to j_5

Job	Workflow template
j_1, j_2	w_1
j_3	w_2
j_4, j_5	w_3

Fig. 5 Time-slot I (length = 7)

Time Sequence	Resources		
	r_1	r_2	r_3
α_1	t_1^1	t_8^5	t_8^4
α_2	t_2^1	t_9^5	t_9^4
α_3	t_3^1	t_4^3	t_{10}^4
α_4	t_{10}^5	t_5^3	t_{11}^4
α_5	t_1^2	t_6^3	t_{11}^5
α_6	t_2^2	t_7^3	
α_7	t_3^2		

Fig. 6 Time-slot II

Resources		
r_1	r_2	r_3
t_1^1	t_1^2	t_8^4
t_2^1	t_2^2	t_9^4
t_3^1	t_3^2	t_4^3
t_5^3	t_{10}^4	t_8^5
t_6^3	t_{11}^4	t_9^5
t_7^3	t_{10}^5	t_{11}^5

jobs as shown in Figs. 3 and 4 resources, r_1, r_2, and r_3 with their executable tasks as in Fig. 2.

In Figs. 5, 6 and 7, three versions of the time-slots are presented. First, the time-slot with length of 7 time units is shown in Fig. 5. Although it is a valid time-slot subjected to the definitions, it is not an optimal time-slot length. Figure 6 shows a time-slot with 6 time units, however, resource r_1 can not be used to execute tasks t_5, t_6, and t_7 as shown in the dash line (the tasks of job j_3). So, Fig. 6 is not a valid time-slot. Last, Fig. 7 shows a valid minimal time-slot for this example, which it is the desirable solution for the problem.

After the required notations are defined, the minimum length time-slot with heterogeneous resources (MLTH) problem can be formulated as follows.

Fig. 7 Time-slot III

Resources		
r_1	r_2	r_3
t_1^1	t_1^2	t_4^3
t_2^1	t_2^2	t_5^3
t_3^1	t_6^3	t_8^4
t_8^5	t_9^5	t_7^3
t_{10}^5	t_3^2	t_9^4
t_{10}^4	t_{11}^4	t_{11}^5

Problem 1 (*MLTH*) Given a set of jobs J in a DDS D that belongs to a set of workflow templates W, find a time-slot S of J on the set of heterogeneous resources Ω that is the length of the time-slot is minimized.

3 A Heuristic Algorithm for the MLTH Problem

In [7], the MLT (minimum length time-slot problem) with homogenous resources is proven as an NP-Complete problem by reducing the problem from the subset sum problem. It can be seen that MLTH is also an NP-Complete problem by reducing the problem from the subset sum problem as well. We omit this proof because of space limitations, however, the same proof approach can be applied. So, we propose an effective heuristic algorithm to schedule the given jobs to the heterogenous resources as follows.

First, we follow an approach presented in [7] to manage the dependency of the given tasks. Thus, the degree of successors of each task is to be determined. Then, the tasks with higher degree of successors are to be executed earlier in order to minimize the waiting time of their successors.

Then, the next issue is to manage the heterogeneity in the DDSs, e.g. how to assign each selected task into the time-slot. For such focused issue, since of the resources are heterogeneous, the resources those can execute the selected task, supposedly t_y^x, are to be considered. Such set of resources is $R = \{r_p | t_y \in \lambda(r_p)\}$. Here, the set of free resources $R' \subseteq R$ is considered, where $R' = \{r_{p'} | S(\alpha_q, r_{p'}) = null\}$. First, for the allocation $S(\alpha_q, r_p) = null, r_p \in R$, we propose that the domain value of S, α_q, must be higher than the time of the predecessors of the task t_y^x in order to guarantee the validity of the workflow template. We also propose to select the free resource with such minimum time to execute the task as soon as possible. Subsequently, we select a single one resource $r_{p''}$ that has the minimum size of its executable set from R', and assign task t_y^x into the slot $S(\alpha_q, r_{p''})$.

The reason for our proposed approach is that, if resources with more heterogeneous, larger executable set, are assigned before less-heterogeneous resources, it can cause the tasks that execute later to have less choice for resource acquisition. Thus, their execution can be delayed, and the whole time-slot length will be longer. As our problem setting also consider task dependencies, the delay can cause more if the successor tasks are effected.

Last, as there can be many tasks with the same degree of successors and many resources with the same size of executable sets, selecting different task or resource can lead to different time-slot assignment. Thus, we utilize the nature of the DDSs by letting all the controllers to determine the local solution differently using the *local algorithm* described above. Subsequently, the time-slot with minimal length will be selected as the final global solution using a simple *interconnect algorithm*.

The details of the proposed algorithms are presented as follows.

3.1 Local Algorithm

The local algorithm for each controller is shown in Algorithm 1. The two major procedures of it are selecting a task for the allocation, and selecting a resource for the selected task as described above.

From the algorithm, first, the degree of successors, *scrDeg*, of each task in all jobs $j_x \in W$ is determined. Also, the set of predecessors, *pdr*, of each task is determined. From the algorithm, the size of *pdr*, denoted as $|pdr_y^x|$, is the degree of predecessors of the task. Then, the task is added to *taskSet* set, which it represents all the tasks in the system.

Subsequently, while the *taskSet* is not empty, the algorithm iterates through the *taskSet*. For each task without predecessor, i.e. *pdrDeg* = 0, it is added to another set, called *useableTask*. This set represents the candidate tasks that can be assigned into the time-slot. Then, the tasks with the highest degree of successors are selected. Though there could be many tasks with the same degree, the controller will randomly select one of them.

After selecting the task, the resource for it has to be decided. It begins with determining the *preAssignedTime* of the task. Formally, *preAssignedTime* of a task is the maximum time of the predecessor of such task, that has been assigned in the time-slot already. Next, the algorithm determines the slots of the resources that can execute the task where *preAssignedTime* + 1 is the beginning time of the valid slot. The *usableSlot* set therefore contains the resources that can execute the task.

Then, the algorithm selects a single slot $S(\alpha_{q'}, r_{p'})$ from *usableSlot* using the heterogeneity of the resource. More specifically, the algorithm selects the slot $S(\alpha_{q'}, r_{p'})$ which the size of the executable set of $r_{p'}$, $|\lambda(r_p)|$, is minimal. If there is more than one slot having the same level of heterogeneity, the algorithm selects the resource randomly.

Finally, the algorithm assigns the selected task to the selected resource. Also, it updates the *assignedTime* of this task, and the length of the time-slot. The *pdrDeg*

Algorithm 1 Local Algorithm

Require: a set of resources Ω of DDS D with λ and a set of jobs J with a set of workflow templates W.

Ensure: a potentially minimal length Time-slot S.

$taskSet \leftarrow \emptyset$, $usableTask \leftarrow \emptyset$, and $timeSlotLength = 0$

for in each job $j_x \in W$ **do**

    Determine the degree of successors of each task t_y in j_x,
    denoted as $scrDeg_y^x$.

    Determine the set of predecessors of each task t_y in j_x
    as pdr_y^x.

    $pdrDeg_y^x = |pdr_y^x|$.

    $taskSet \leftarrow taskSet \cup \{t_y^x\}$ where t_y^x is t_y from j_x

end for

while $taskSet \neq \emptyset$ **do**

    $usableTask \leftarrow \emptyset$

    for each task $t_y^x \in taskSet$ **do**

        if $pdrDeg_y^x = 0$ **then**

            $usableTask \leftarrow usableTask \cup \{t_y^x\}$

        end if

    end for

    Determine the set of maximum-$scrDeg$ tasks from
    $usableTask$, denoted as max_scrDeg.

    Select task t_h^g from max_scrDeg randomly.

    Determine $preAssignedTime$ which is
    $max(\{assignedTime(t)|t$ is the predecessor of $t_h^g\})$,
    if t_h^g is the start task, $preAssignedTime = 0$.

    $usableSlot = \emptyset$

    while $usableSlot = \emptyset$ **do**

        $usableSlot = \{S(\alpha_q, r_p)|S(\alpha_q, r_p) = null,$
        $t_h^g \in \lambda(r_p), \exists r_p \in \Omega,$
        and $\alpha_q = preAssignedTime + 1\}$

        $preAssignedTime = preAssignedTime + 1$

    end while

    Select slot $S(\alpha_{q'}, r_{p'})$ such that $|\lambda(r_{p'})|$ is the minimum
    from all elements in $usableSlot$, if there is more than
    one slot, select a slot randomly.

    $S(\alpha_{q'}, r_{p'}) = t_h^g$

    $assignedTime(t_h^g) = \alpha_{q'}$

    $taskSet = taskSet - \{t_h^g\}$

    if $\alpha_{q'} > timeSlotLength$ **then**

        $timeSlotLength = \alpha_{q'}$

    end if

    for each successor $t_{h'}^{g'}$ of t_h^g **do**

        $pdrDeg_{h'}^{g'} = pdrDeg_{h'}^{g'} - 1$

    end for

end while

return S and $timeSlotLength$

---

of successor of the assigned task is reduced by one. Such algorithm keeps repeating this described procedure until all the tasks are assigned to the time-slot.

The cost to resolve the MLTH problem using Algorithm 1 is $O(n^3 m)$ where n is the number of all tasks, and m is the number of all resources. The main cost comes from the *usableSlot* determination, i.e. the set of slots that can assign the selected task into it. In each task, it takes $O(n^2 m)$ to determine the *usableSlot*. Since, such computing is required until all tasks are completely assigned, so, the cost is $O(n^3 m)$.

For the sake of clarity, we present an example to illustrate our local algorithm execution as follows. Let the set of jobs are given as shown in Fig. 4, and the set of resources and their executable set are as shown in Fig. 2. First, all of the start tasks from all jobs are considered as *useableTask* since their $pdrDeg = 0$. Then, the algorithm selects task t_4^3 from *useableTask* because it has the maximum degree of successors. As task t_4 of job j_3 is either in $\lambda(r_2)$ and $\lambda(r_3)$, so *usableSlot* = $\{S(\alpha_1, r_2), S(\alpha_1, r_3)\}$. Since t_4^3 is the start task, its $preAssignTime = 0$, and $S(\alpha_1, r_2)$ and $S(\alpha_1, r_3)$ is *null*, both slots are in *usableSlot* set. Finally, the algorithm selects $S(\alpha_1, r_3)$ to execute task t_4^3 because $|\lambda(r_3)| < |\lambda(r_2)|$. It also updates the *assignedTime* of t_4^3, *timeSlotLength* and *pdrDeg* of successors of t_4^3. The algorithm repeats all of the procedures until $taskSet = \emptyset$. Figure 7 is the solution from this running example.

3.2 Interconnect Algorithm

Once the local solutions have been computed by all the controllers using Algorithm 1, the global final solution is to be determined. It can be done simply by comparing the local solution from each controller, i.e. the time-slot length information. Subsequently, the minimal global time-slot length is determined, and such assignment is ready to be executed.

4 Evaluation

In this section, we present the experiment results to evaluate our proposed work.

4.1 Simulation Setup

To evaluate our work, a workflow synthetic-data generator, that generates the workflow template using the number of workflow templates, number of minimal and maximal tasks per jobs, and number of jobs as the inputs, is implemented.

In the experiments, the number of resources, the number of controllers, and the number of workflow tempts are fixed at 100 resources, 50 controllers, and 10 workflow templates respectively. The number of tasks of each workflow is fixed between 45–50 tasks. The degree of successors of each node is set between 0–5. The number of tasks in the longest path of each workflow template is set between 15–35 tasks. The workflow template for each job is selected uniform randomly. In order to guarantee that all jobs can be executed, in each experiment, 10 % of the resources are set to be able to execute all tasks from all workflow templates.

The experiment results of our proposed work are compared with a baseline algorithm in [7]. Such algorithm has demonstrated for its efficiency and effectiveness, i.e. its polynomial-time complexity and the solutions which are close to the theoretical results respectively. Both algorithms are implemented using Java SE 7. The experiments are conducted on an Intel Core 2 Duo 2.4 GHz with 4 GB memory running Mac OS X.

4.2 Result

First, to evaluate the impact of the heterogeneity on the performance of our algorithm, the size of the executable sets is varied from 10 to 100 % of the number of tasks from all workflow templates, while the number of jobs is fixed to 500 jobs.

Figure 8 shows the experimental results. It can be seen that when the size of executable set of each resource is small (10–30 %), the time-slot lengths from both

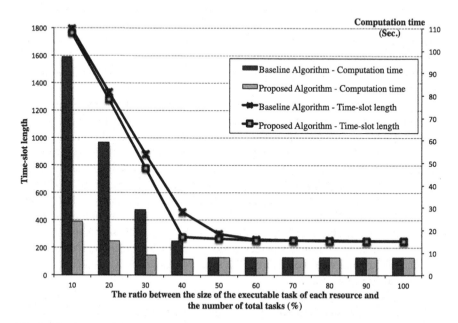

Fig. 8 The time-slot length and the computation time when the heterogenous level is varied

algorithms are quite large. However, the computation time of the proposed algorithm is very low comparing with the computation time of the algorithm in [7]. It is because the algorithm in [7] will specify the resource firstly, then it select a task form all available tasks to assign into the time-slot at this resource. If there is no task to be executed by this resource, the algorithm will select the next resource and finds a task for the assignment again. This can enormous delay the assignment in the heterogenous environment. Meanwhile, our proposed work will select the resource based on the executable set, which is also first available for the task (its $\alpha_q = preAssignedTime + 1$). So, the execution time of our proposed work is much less than the baseline algorithm. When the size of the executable set is set at 30–50 %, the time-slot lengths from both algorithms are very different. The most different point is when the size of the executable set is fixed at 40 %, in which the time-slot length from proposed algorithm is 275 time units, while the time-slot length from [7] is 457 time units. The reason behind this is that the proposed algorithm considers the heterogeneity of the resources. Specifically, the resources with the small executable set are always allocated first. So, we preserve the resources with the larger executable set in the earlier phase of computing. When the algorithm finds the resources for the selected task later on, such task can be assigned earlier. In the other words, the delay due to waiting for the available executable resource is less. This gives more advantage particularly for the successor-task as discussed in the previous section. With this reasoning, the small time-slot length of the proposed algorithm can be achieved.

When the size of the executable sets are more than 50 %, the performance of both algorithms are similar. The reason is that each resource can execute many tasks from many workflow templates at these sizes, so, it has small number of *null* slots in both algorithms. Then, the time-slot lengths of both algorithms are too low. With the same reason, any selected task can be always assigned to the proper resource, this makes the computation time of the algorithm in [7] close to the computation time of the proposed algorithm.

After the impact of the task scheduling using the size of the executable sets in the heterogeneous resources has been evaluated, we then evaluate the impact of the number of jobs, or the scalability. The number of the jobs is varied from 50 to 500 jobs. In this experiment, we randomly generate the resources with the size of their executable sets, the heterogeneity of the resources, at a moderate level, i.e. 35–45 % of the number of total tasks from all workflow templates. The reason is that too large executable sets can cause the resources to be able to execute too many tasks from all workflow templates. In the other words, it causes the heterogeneity of the resources undistinguished. On the other hand, if the size of the executable sets is too small, it causes the resources to be able to execute too few tasks. So, the amount of delayed tasks can be extreme large due to waiting for available resources. This can be extreme cases in real-life applications.

Figure 9 shows the performance in terms of the time-slot length and the execution time. First, we consider the solutions from the experiments, i.e. minimal time-slot lengths. Figure 9 shows that the time-slot length from the proposed algorithm is obviously less than the time-slot length from the algorithm in [7]. And, when the

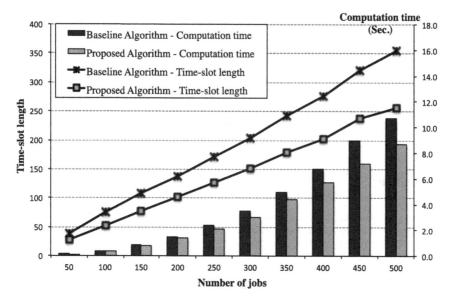

Fig. 9 The minimum length of time-slot and the computation time when the number of jobs is varied

number of jobs is increased, the difference of the time-slot length of two algorithms are increased. When considering the execution time, when the number of jobs is increased, the proposed algorithm is much efficient. The reason behind this is the complexity of the algorithm in [7] is $O(n^3m^2)$ meanwhile the complexity of the proposed algorithm is only $O(n^3m)$. When the execution time with the effectiveness of the solutions is considered, the proposed algorithm is highly effective. Even when the number of jobs is set at 500, our algorithm takes only 8.7 s to determine the time-slot which is very small. Thus, not only the algorithm is effective, but also it is efficient in decentralized distributed system with the heterogeneous resources.

5 Conclusion and Future Work

In this paper, we have addressed a scheduling problem in decentralized distributed systems with heterogeneous resources for the jobs with dependency, so called MLTH. As it is an NP-Complete problem, thus, a heuristic algorithm is proposed instead of aiming at the exact solution. Our proposed algorithm is based on the observation that the length of the schedule, time-slot length, can be reduced, if the resources with less heterogeneity is assigned to the tasks earlier. This not only can generate a smaller time-slot length, but the less execution time to generate the schedule can also be achieved. In order to evaluate the proposed work, the experiment results are presented. The results show that our approach is highly effective and also efficient, particularly, when the heterogeneity is at moderate level and the number of jobs is large. In the future work, we intend to investigate the approximation approach which can guarantee the quality of the solution. Moreover, we intend to address the similar problem with different scheduling objectives.

Acknowledgments The work is partially supported by Graduate School of Chiang Mai University. The authors would like to thank the colleagues at Data Engineering and Network Laboratory, Faculty of Engineering, Chiang Mai University for their support.

References

1. Kondo, D., Andrzejak, A., Anderson, D.P.: On correlated availability in internet-distributed systems. In: Proceedings of the 9th IEEE/ACM International Conference on Grid Computing, pp. 276–283. Washington, DC, USA (2008)
2. Lai, K., Huberman, B.A., Fine, L.R.: Tycoon: A distributed market-based resource allocation system. Comput. Res. Repos. cs.DC/0404013 (2004)
3. Mainland, G., Parkes, D.C., Welsh, M.: Decentralized, adaptive resource allocation for sensor networks. In: Proceedings of the 2nd conference on Symposium on Networked Systems Design and Implementation, Vol. 2, pp. 315–328. Berkeley, CA, USA (2005)
4. Masuishi, T., Kuriyama, H., Oki, Y., Mori, K.: Autonomous decentralized resource allocation for tracking dynamic load change. In: Proceedings of the International Symposium on Autonomous Decentralized Systems, pp. 277–283 (2005)
5. Pathan, AsK, Pathan, M., Lee, H.Y.: Advancements in Distributed Computing and Internet Technologies: Trends and Issues, 1st edn. Information Science Reference - Imprint of: IGI Publishing, Hershey, PA (2011)
6. Sotiriadis, S., Bessis, N., Xhafa, F., Antonopoulos, N.: From meta-computing to interoperable infrastructures: A review of meta-schedulers for hpc, grid and cloud. Advanced Information Networking and Applications, International Conference on 0, 874–883 (2012). http://doi.ieeecomputersociety.org/10.1109/AINA.2012.15
7. Tantitharanukul, N., Natwichai, J., Boonma, P.: Workflow-based composite job scheduling for decentralized distributed systems. In: Proceedings of the Sixteenth International Conference on Network-Based Information Systems (NBiS), pp. 583–588 (2013)
8. Tsamoura, E., Gounaris, A., Manolopoulos, Y.: Decentralized execution of linear workflows over web services. Futur. Gener. Comput. Syst. **27**(3), 290–291 (2011)

Novel Data Integrity Verification Schemes in Cloud Storage

Thanh Cuong Nguyen, Wenfeng Shen, Zhaokai Luo, Zhou Lei
and Weimin Xu

Abstract In order to persuade users of widely using cloud storage, one critical
challenge that should be solved is finding way to determine whether data has been
illegally modified on the cloud server or not. The topic has although been addressed in
several works, there is lack of scheme to meet all the demand of supporting dynamic
operations, public verification, less computation etc. This paper surveys the related
results that has been done and proposes two alternative schemes, called DIV-I and
DIV-II. Compared to S-PDP introduced by Ateniese et al., both DIV-I and DIV-II
use less time to generate tags and verify. In addition, the proposed schemes fully
support dynamic operations as well as public verification.

Keywords Cloud storage · Data integrity · Dynamic data

1 Introduction

When using cloud storage, user enjoys many prominent characteristics such as
accessing data anytime and anywhere, being released from arduous work of main-
taining hardware and software, etc. Those cutting edges promote the wide adaptation
of cloud storage in practice. However, many critical security challenges emerge due
to user's limited control over his data, which totally differentiate to traditional storage approach. One of those challenges is how to determine the intactness of data is
ensured or not. Notice that the data can be damaged by many reasons from malicious
attack to hardware failure, and the cloud storage providers (CSP) may hide that to
hold their reputation. Hence, establishing a scheme to verify the data's intactness is
a key requirement.

T.C. Nguyen (✉) · W. Shen · Z. Luo · Z. Lei · W. Xu
School of Computer Engineering and Science,
Shanghai University, Rm. 809, Computer Building, 333 Nanchen Road, Shanghai 200444, China
e-mail: ruan3@shu.edu.cn

W. Shen
e-mail: wfshen@mail.shu.edu.cn

© Springer International Publishing Switzerland 2015
R. Lee (ed.), *Computer and Information Science*, Studies in Computational
Intelligence 566, DOI 10.1007/978-3-319-10509-3_9

Such a verification scheme should have the following features, as many as possible:

Support public verification: The scheme should allow not only data owner but also anyone authorized in the cloud system to verify the data integrity. This is especially important because when outsourcing data, user sometime wants to share his data to his friends or partners. They need to have ability to make sure that the data they retrieve are not illegally modified. Additionally, user, who is not online frequently, can delegate the verification task to a third party auditor (TPA). The TPA will send the verification request periodically to CSP to ensure that any data corruption is detected in time. TPA can even play more essential role in evaluating the quality of service of CSP. Any CSP that has poor profile in terms of ensuring the innocence of the data it stores will probably lose their users.

Unlimited verification time: For each verification request, auditor needs to send a challenge to CSP, and CSP is supposed to return appropriate result corresponding to the challenge. If the scheme allows only limited verification time, some challenges must be repeated after all of them has been used. That brings CSP a wonderful chance to answer with a deceived return if it stores all of previous response corresponding to each challenge.

No data leakage: During the verification process, the data owner may not want to reveal any data to TPA. Hence, the scheme should protect the data privacy against TPA no matter how many tuple (challenge, response) is collected.

Support data dynamic operations: The data owner sometime wants to modify his file such as deleting part of the file or inserting more data somewhere in the file. In this case, the file's tags need to be recomputed. Thus, the scheme should be able to update the tags with lowest cost.

This paper proposes two schemes that address all above concerns. The proposed schemes fully support data dynamic operations as well as public verification without data leakage. Related demonstration and experiment are carried out to prove their correctness and outstanding performance compared to other schemes.

The rest of this paper is organized as follows. Section 2 reviews some related work. Section 3 introduces two novel schemes DIV-I and DIV-II in detail. Section 4 presents experiment results and analysis. And the last section makes conclusion.

2 Related Work

Many outstanding work has been done to provide judgment scheme for data integrity on cloud storage. Some of them generates limited number of tuple (challenge, response) before outsourcing data. For example, Juels and Kaliski introduced POR scheme [1] which embedded a number of special blocks, called 'sentinels', among file blocks, and the verifier releases one sentinel's position for each challenge. Aravan and Ashutosh [2] selected from each block a certain number of bits to compose its verification proof. This scheme has been improved in [3] to support data dynamic operations, but limited verification time still remains. Ateniese et al. [4] used cryptographic hash function to generate verification proof. This scheme supports block update, deletion and append. However, block insertion anywhere in the file is not allowed. Chaoling li et al. [5] tried to improve scheme of [4] to support block insertion by using SN-BN table. However, the authors did not realizes that correctness of this scheme was not ensured if the integrity of SN-BN table was not guaranteed. Noticeably, if the CSP stores full set of tuples, all above schemes are not reliable any more.

On the other hand, some solutions based on RSA and Diffie-Hellman assumptions support dynamic operations such as Deswarte and Quisquater [6] and Sebe et al. [7]. However, those schemes do not support public verification. Some another solutions using homomorphic authentication like Shacham and Waters [8] and Ateniese et al. [9] allow unlimited public verification time, but may lead to data leakage. We notice that S-PDP introduced in [9] can be improved by masking the CSP's response with a random number to prevent data leakage. Erway et al. [10] proposed another model based on Ateniese et al.'s [9] model to better support dynamic operations. Liu et al. [11] improved Erway et al.'s [10] model to reduce computational and communication overhead. Neither of them allows public verification. Wang et al. [12] took advantage of both homomorphic authentication and Merkle hash tree [13] to allow both unlimited public verification and dynamic operations. However, data privacy was not considered in this work. Zhu [14] introduced a scheme based on Diffie-Hellman assumption [15] and bilinear pairings. Although the scheme supports public verification, dynamic operations were not addressed in this work.

3 The Novel Schemes

Let p, q, be two large primes and $N = p * q$ be an RSA modulus. Let $\varphi(N) = (p - 1) * (q - 1)$ be the Euler function of N, and d, e are two big integers satisfy $d * e \equiv 1 \mod \varphi(N)$. Let l is a security parameter, assume that $|d| \geq l, |e| \geq l$, ($|d|$, $|e|$ are bit-length of d and e respectively). N and e are made public while p, q, $\varphi(N)$, d are only known by the data owner. Additionally, let g be an element with high order in $\mathbb{Z}_N^*$ and g is coprime to N. g is also made publicly known.

We suppose that data owner has a file, which includes n blocks, each block has bits, needs to be outsourced. Hence, the file size is $n * s_b$ bits. In this paper, a tag is calculated for each block as its authentication data. The ith block is denoted by b_i and its tag is denoted by T_i.

3.1 DIV-I Scheme

Here, we propose a scheme which is similar to S-PDP but performances better in term of tag generation and integrity verification. We call the scheme DIV-I. The scheme includes four functions as follows:

GenKey$(l_1, l_2) \rightarrow (pk, sk)$: generates a public key $pk = (N, e, v, r, g)$ and private key $sk = (\varphi(N), d, \alpha)$, where r, α are two random numbers and $v = g^\alpha \bmod N$. Let l_1, l_2 are two security parameter, $r \leftarrow \{0, 1\}^{l_1}, \alpha \leftarrow \{0, 1\}^{l_2}$

GenTag$(pk, sk, i, b_i) \rightarrow T_i$: Let $h_i = H(r\|i)$, where H is responsible to compute the hash value and convert to big integer. Data owner computes $T_i = g^{(\alpha h_i + b_i)*d} \bmod N$ and sends $\{(b_i, T_i)\}_{0 \le i < n}$ to CSP.

GenProof$(pk, F, chal) \rightarrow (T, B)$: Verifier sends challenge to CSP in a form of $(g_s, c, \{(i_j, c_j)\}_{1 \le j \le c})$, where $g_s = g^s \bmod N$, s is a random number, c is the number of blocks that compose the response, (i_j, c_j) is those blocks' index and coefficient, correspondingly. CSP responds two values:

$$T = \Pi_{j=1}^c T_{i_j}^{c_j} \bmod N \text{ and } B = g_s^{\sum_{j=1}^c c_j b_{i_j}} \bmod N$$

VenProof$(pk, chal, T, B) \rightarrow \{"Y", "N"\}$: Verifier computes $h = v^{\sum_{j=1}^c c_j h_{i_j}} \bmod N$ and check the condition $T^{e*s} = B * h^s \bmod N$. If the condition is satisfied, then returns "Y" indicates no data corruption detected. Otherwise, returns "N" indicates Data corruption detected.

Correctness: In case of the intact of all b_{i_j} and T_{i_j} is ensured, we have $T^{e*s} = g^{s*\sum_{j=1}^c (c_j b_{i_j} + \alpha c_j h_{i_j})} \bmod N$

$$= g^{s*\sum_{j=1}^c c_j b_{i_j}} * g^{\alpha s * \sum_{j=1}^c c_j h_{i_j}} \bmod N$$

$$= B * (g^{\alpha * \sum_{j=1}^c c_j h_{i_j}})^s \bmod N$$

$$= B * h^s \bmod N$$

Robustness: Assuming that the factorization, RSA and Diffie-Hellman problem are difficult over $\mathbb{Z}_N^*$, CSP successfully pass the challenge for DIV-I scheme if and only if the intact of all blocks and tags participate in the response is ensured.

Proof Suppose that some of blocks participate in the response are corrupted. We prove that if CSP successfully pass the challenge, there is method to break RSA problem. That means with any integer z, it is able to find a value w that satisfies $w^e = z \bmod N$ without knowing d. The construction of this method is describes as follows.

In the above construction of DIV-I, let $g = z$. We assume that there are k corrupted blocks in total c blocks compose B, and w.l.o.g they are $b'_{i_1}, b'_{i_2}, \ldots, b'_{i_k}$. CPS responds to verifier a tuple (T', B'), where $B' = g_s^{b'} \bmod N$. Without knowing s, in order to pass the challenge, CSP needs to ensure that $T'^e = g^{b'} * h \bmod N$ (2). Let $u = \sum_{j=1}^c, c_j h_{i_j}$ (2) $\Leftrightarrow T'^e = z^{b'} * z^{\alpha u} \bmod N$. Because α is unknown to CSP and we can choose e as a large prime, thus w.l.o.g we assume that $gcd(e, b' + \alpha u) = 1$. Thus the extended Euclidian algorithm can be used to find out x and y such that $ex + (b' + \alpha u)y = 1$. Let $w = z^x * T'^y$, we have $w^e = z^{ex} * T'^{ey} = z^{ex} * z^{(b' + \alpha u)y} = z$. That means the RSA problem has been break.

Update: If data owner inserts a block b at the position pos, all the tag of blocks from pos to n need to be updated. The update process is summarized as follow:

- CSP sends $\{T_i\}_{pos \leq i \leq n}$ to data owner.
- Data owner computes new tags: $T'_{i+1} = T_i * g^{\alpha(h_{i+1}-h_i)*d} \bmod N$
- Data owner sends $\{T'_{i+1}\}_{pos \leq i \leq n}$ to CSP.

The maximum communication overhead is $(2n+1)*|N|+s_b$ when $pos = 0$ and the minimum one is $|N| + s_b$ when $pos = n + 1$ ($|N|$ is bit-length of N). The deletion process is carried out similarly.

Storage: pk and sk need a storage space of $5|N|+l_1+l_2$ bits (g is ignored because we can set g as small integer like 2, 3, 5). Extra storage to keep all tags on CSP is $*|N|$. Thus, the ratio of extra storage and file size is $|N|/s_b$. For example, in case of a 4GB file includes $n = 1,000,000$ blocks, each block is 4 KB-size, $|N| = 1024$ bits and $l_1 = l_2 = 128$ bits, CSP need 122 MB extra storage, pk and sk need 5.25 KB.

3.2 DIV-II Scheme

We try to reduce power and multiplication computation compared to above schemes to hopefully lessen the computation cost. The new scheme, called DIV-II, is described as follow:

GenKey$(l_1, l_2) \rightarrow (pk, sk)$: generates a public key $pk = (N, e, r, g, v)$ and private key $sk = (\varphi(N), d, \alpha, u)$, where r, u are random numbers and $v = g^\alpha \bmod N$. Let l_1, l_2, are two security parameter, $r \leftarrow \{0, 1\}^{l_1}, u \leftarrow \{0, 1\}^{l_1}, \alpha \leftarrow \{0, 1\}^{l_2}$.

GenTag$(pk, sk, i, b_i) \rightarrow (T_i, \beta_i)$: Let $h_i = H_1(r||i), h'_i = H_2(u||i)$ and $\beta_i = g^{h'_i} \bmod N$, where H_1, H_2 are two functions that compute the hash value of string and convert to big integer. Data owner generates tag for all block using the formula $T_i = \alpha * b_i + d * h_i + h'_i \bmod \varphi(N)$ mod then sends $\{(b_i, T_i, \beta_i)\}_{0 \leq i \leq n}$ to CSP.

GenProof$(pk, F, chal) \rightarrow (T, B)$: Verifier sends challenge to CSP in a form of $(c, \{(i_j, c_j)\}_{1 \leq j \leq c})$, where is the number of blocks that compose the response, (i_j, c_j) is those blocks' index and coefficient, correspondingly. CSP responds three values:

$$T = g^{\sum_{j=1}^{c} c_j T_{i_j}} \bmod N, \ B = v^{\sum_{j=1}^{c} c_j b_{i_j}} * \Pi_{j=1}^{c} \beta_{i_j}^{c_j} \bmod N.$$

VerProof$(pk, chal, T, B) \rightarrow \{"Y", "N"\}$: Verifier computes $h = g^{\sum_{j=1}^{c} c_j h_{i_j}}$ $\bmod N$ and check the condition $T^e = B^e * h \bmod N$. If the condition is satisfied, then return "Y" means "No data corruption detected". Otherwise, return "Y" means "Data corruption detected".

Correctness: In case of the intact of all b_{i_j}, T_{i_j} and β_{i_j} is ensured, we have:

$$T^e = g^{e*\sum_{j=1}^{c} c_j T_{i_j}} \bmod N$$

$$= g^{e * \sum_{j=1}^{c} c_j (\alpha * b_{i_j} + d * h_{i_j} + h'_{i_j})} \bmod N$$

$$= g^{e * \alpha * \sum_{j=1}^{c} c_j b_{i_j}} * g^{\sum_{j=1}^{c} c_j h'_{i_j}} * g^{\sum_{j=1}^{c} c_j h_{i_j}} \bmod N$$

$$= B^{e} * h \bmod N$$

Robustness: Assuming that the factorization, RSA and Diffie-Hellman problem are difficult over $\mathbb{Z}_N^*$, CSP successfully pass the challenge for DIV-II scheme if and only if the intact of all blocks and tags participate in the response is ensured.

Proof We prove that if CSP's response to challenge can successfully pass the verification while some block has been corrupted, there is scheme to break RSA problem. For instance, for a particular z, we are able to find out a value w that satisfies $W^e = z \bmod N$.

In the above scheme, let $g = z$. Suppose that CSP responds T and B. Let $u = \sum_{j=1}^{c} c_j h_{i_j}$. If the response passes the verification, the equation $T^e = B^e * z^u \bmod N$ (3) should be established. (3)$\Leftrightarrow$ $(TB^{-1})^e = z^u \bmod N$. We assume w.l.o.g that $gcd(e, u) = 1$. Thus we can find out x and y such that $ex + uy = 1$. Let $w = z^x * (TB^{-1})^y$, we have $w^e = z^{ex} * (TB^{-1})^{ey} = z^{ex} * z^{uy} = z$. That means the RSA problem has been break.

Update: In DIV-II, when data owner inserts a block at the position , all the tags need to be updated. The update process is summarized as follow:

- CSP sends $\{T_i\}_{0 \le i \le n}$ to data owner.
- Data owner chooses a random number u'. Let $h''_i = H_2(u' \| i)$
- Data owner computes new tags:

$$T'_i = T_i + h''_i - h'_i \bmod \varphi(N) \text{ if } 0 \le i < pos$$

$$T'_{i+1} = T_i + d * (h_{i+1} + h_i) + h''_{i+1} - h'_i \bmod \varphi(N) \text{ if } pos \le i < n$$

- Data owner sends $\{T'_i\}_{0 \le i \le n}$ to CSP.

Like the previous scheme, the update process of DIV-II is without downloading the block's data. The communication overhead is $(2n+1) * |N| + s_b$. However, if new block is appended, no tag need to be updated. Thus, the communication overhead is only $|N| + b_s$.

Unlike DIV-I scheme, when a new block inserted, DIV-II requires all tag to be recomputed. If tag of blocks prior position are not recomputed, there's no need to choose u'. Hence, $T'_{i+1} = \alpha * b_i + d * h_{i+1} + h'_{i+1} \bmod \varphi(N)$, $pos \le i < n$. Because CSP knows $T_{i+1} = \alpha * b_{i+1} + d * h_{i+1} + h'_{i+1} \bmod \varphi(N)$, $pos \le i < n$ it can compute $T'_{i+1} - T_{i+1} = \alpha * (b_i - b_{i+1}) \bmod \varphi(N)$. Thus, CSP can find out α as well as $\varphi(N)$, and it will be able to compute all the tag.

Storage: compared to DIV-I, the ratio of extra storage on CSP and file size is $2|N|/s_b$.

In both DIV-I and DIV-II, in order to avoid storing many r, we can set this value to each file's unique sequence provided by CSP. Additionally, if we use two seeds, one used to generate i_j and the other used to generate c_j, just like S-PDP scheme, the challenge communication overhead can be reduced from $O(c)$ to $O(1)$. Besides that, unlike some previous schemes, verifier does not receive a linear combination of data blocks, hence data privacy is preserved.

The deletion process is similar to insertion.

4 Experiment and Analysis

4.1 Experiment Environment and Assumptions

We code C++ programs to run on a computer with Intel(R) Core(TM) 2 Quad CPU Q8200@2.33 GHz, 2.34 GHz and RAM of 2 GB to test the performances. We want to compare the computation time of two proposed schemes with S-PDP scheme, thus we just need one computer to run all functions including GenKey, GenTag, GenProof, VerProof. We do not use two seeds to generate i_j and c_j. Those values are included in the challenge. Additionally, we use MD5 to compute all hash value. Algorithms use NTL library [16] for doing big number calculation. Each case in our experiment is repeated 10 times and the mean value is used for further analysis.

4.2 Results and Analysis

First of all, we test the performance of three schemes (S-PDP, DIV-I, DIV-II) for 1 MB file in different cases corresponding to different value of block size. As can be seen from Fig. 1, the tag generation time of all three scheme is inversely proportional to block size. Additionally, the time of S-PDP is the biggest and the time of DIV-II is significantly less than that of the others. Interestingly, Fig. 2 shows that the server computation time is absolutely the same in three scheme independently to block size. Moreover, the time decreases when block size is smaller than 4 KB, reaches the bottom when block size is equal to 4 KB, and increases with larger block. Table 1 presents that the verification time of S-PDP is inversely proportional to block size while the time of other schemes just slightly decreases when block size varies from 1 to 64 KB. Furthermore, the time of our two schemes is less than that of S-PDP in all cases.

On the other hand, Fig. 3 depicts inversely proportional relation of maximum insertion time and block size for two novel schemes. Notably, the time of DIV-I is the bigger than that of DIV-I. In Fig. 4, however, the minimum insertion time of DIV-II is the bigger. In fact, the maximum and minimum insertion time of DIV-II is the same because all blocks need to be updated in both situations.

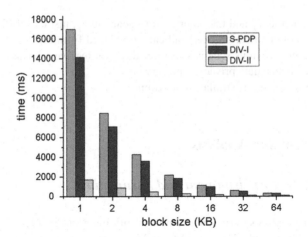

Fig. 1 Tag generation time for 1 MB file

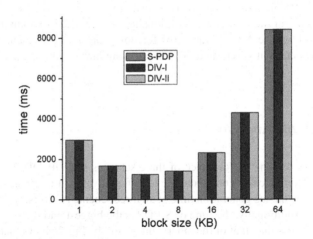

Fig. 2 Server computation time for 1 MB file

Table 1 Verification time for 1 MB file

Block size (KB)	S-PDP (ms)	DIV-I (ms)	DIV-II (ms)
1	2602	32	43
2	1298	27	39
4	659	27	37
8	337	25	37
16	180	25	37
32	97	24	36
64	58	24	36

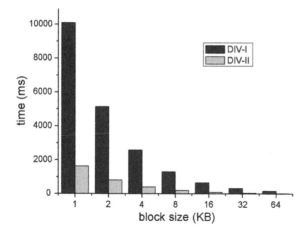

Fig. 3 Maximum insertion time for 1 MB file

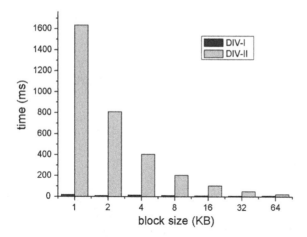

Fig. 4 Minimum insertion time for 1 MB file

Next, we set block size to 4 KB and test the performance of three scheme with different file size. Note that when block size is constant, number of block is directly proportional to file size. Figure 5 shows the directly proportional relation of server computation time, which is the same for three schemes, and file size. However, in case of verification, as can be seen from Table 2, the time of our two schemes only lightly rises while that of S-PDP still directly proportional to file size and the time of S-PDP is always bigger than that of two proposed schemes in all cases.

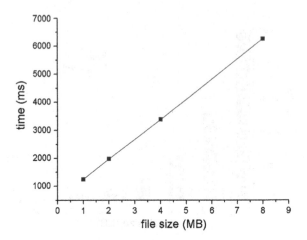

Fig. 5 Server computation time for 4 KB block size

Table 2 Verification time in case of 4 KB block size

File size (MB)	S-PDP (ms)	DIV-I (ms)	DIV-II (ms)
1	659	27	37
2	1311	28	40
4	2593	30	43
8	5160	37	48

5 Conclusion and Future Work

This paper proposes two novel schemes to verify the data integrity in cloud storage. Those schemes allow unlimited verification time and third party verification. Moreover, those support public verification and do not introduce any data leakages. Compare to S-PDP, those schemes need fewer computation time to generate tags and verify data integrity. Additionally, DIV-II needs more extra storage on CSP and dramatically decreases tag generation time compared to DIV-I. However, the verification time of DIV-II is slightly bigger than that of DIV-I. Interestingly, we notice that the two novel scheme have potential to combine with error-correcting like in POR and with spot checking referred in [9] to obtain better performance. This idea should be addressed in the future work.

Acknowledgments This study was supported by National High-tech R&D Program of China (Grant NO. 2009AA012201), the Shanghai Leading Academic Discipline Project (Project No.J50103), and the Innovation Project of Shanghai University.

References

1. Juels, A., Kaliski, B.S.Jr: PORs : proofs of retrievability for large files. In: Proceedings of the 2007 ACM Conference on Computer and Communications Security, CCS 2007, pp. 584–597 (2007)
2. Sravan Kumar, R., Saxena, A.: Data integrity proofs in cloud storage. In: 2011 3rd International Conference on Communication Systems and Networks (COMSNETS), pp.1–4 (2011)
3. Nguyen, T.C., Shen, W., et al.: A probabilistic integrity checking scheme for dynamic data in untrusted cloud storage. In: 12th IEEE/ACIS International Conference on Computer and Information Science (ICIS 2013), Niigata, Japan, pp. 179–183 (2013)
4. Ateniese, G., Pietro Di, R., Mancini, L.V., Tsudik, G.: Scalable and efficient provable data possession. In: Proceedings of the 4th iIternational Conference on Security and Privacy in Communication Networks - SecureComm 08, pp. 1–10 (2008)
5. Li, C., Chen, Y., Tan, P., Yang, G.: An efficient provable data possession scheme with data dynamics. International Conference on Computer Science & Service System (CSSS), pp. 706–710 (2012)
6. Deswarte, Y., Quisquater, J.: Remote integrity checking. IFIP Int. Fed. Inf. Process. **140**, 1–11 (2004)
7. Sebe, F., Domingo-Ferrer, J., Martinez-Balleste, A., Deswarte, Y.: Efficient remote data possession checking in critical information infrastructures. IEEE Trans. Knowl. Data Eng. **20**(8), 1034–1038 (2008)
8. Shacham, H., Waters, B.: Compact proofs of retrievability. In: ASIACRYPT '08 Proceedings of the 14th International Conference on the Theory and Application of Cryptology and Information Security: Advances in Cryptology, pp. 90–107 (2008)
9. Ateniese, G., Burns, R., Curtmola, R., Herring, J., Kissner, L., Peterson, Z., Song, D.: Provable data possession at untrusted stores. In: Proceedings of the 14th ACM Conference on Computer and Communications Security - CCS '07, pp. 598–609 (2007)
10. Erway, C.C., Küpçü, A., Papamanthou, C., Tamassia, R.: Dynamic provable data possession. In: Proceedings of the 16th ACM conference on Computer and Communications Security - CCS '09, pp. 213–234 (2009)
11. Liu, F., Gu, D., Lu, H., Chris, C.: An improved dynamic provable data possession model. In: IEEE International Conference on Cloud Computing and Intelligence Systems (CCIS), pp. 290–295 (2011)
12. Wang, Q., Wang, C., Li, J., Ren, K., Lou, W.: Enabling public verifiability and data dynamics for storage security in cloud computing. In: Proceedings of the 14th European Symposium on Research in Computer Security, ESORICS 2009, pp. 355–370 (2009)
13. Merkle, R.C.: Protocols for public key cryptosystems. In: Proceedings of IEEE Symposium on Security and Privacy'80, pp. 122–133 (1980)
14. Zhu, Y., Hu, H., Ahn, G., Yau, S.S.: Efficient audit service outsourcing for data integrity in clouds. J. Syst. Softw. **85**(5), 1083–1095 (2012)
15. Dan, B.: The decision Diffie-Hellman problem. In: Proceedings of the 3rd Algorithmic Number Theory Symposium. Lecture Notes in Computer Science 1423, pp. 48–63 (1998)
16. Shoup, V.: NTL: a library for doing number theory. http://www.shoup.net/ntl/. Accessed 29 Dec 2013

Generation of Assurance Cases for Medical Devices

Chung-Ling Lin and Wuwei Shen

Abstract In safety critical systems, the manufacturers should provide compelling and comprehensible arguments to demonstrate that their system is well designed and safety of the system to the public is guaranteed. These arguments are usually represented by an assurance case. However, one of challenging issues facing the safety critical industry is how to produce an assurance case that provides a set of well-structured arguments connecting safety requirements and a body of evidence produced during software development. In this paper, we take the medical systems industry into account to illustrate how an assurance case can be generated when a software process is employed. In particularly, we consider the Generic Insulin infusion Pump (GIIP) to show how an assurance case can be produced via during a popular software development process, called Rational Unified Process (RUP).

Keywords Assurance cases · UML profile · OCL · Medical device software · Safety critical system

1 Introduction

One of the most challenging issues facing the safety critical industry is how to develop an assurance case providing a compelling and comprehensible argument to demonstrate that a safety critical system is well designed so its safety is guaranteed when it is in use. In safety critical domains, there are many international standards where many safety-related requirements are specified. A well-designed assurance case should successfully link the evidence to the specific safety objective of the system in a convincing way. But, how to produce such an assurance case has become an important issue. Many manufacturers have found the generation of an assurance case after a software system has been developed is quite time consuming

C.-L. Lin (✉) · W. Shen
Department of Computer Science, Western Michigan University,
Kalamazoo, MI, USA
e-mail: chung-ling.lin@wmich.edu

W. Shen
e-mail: wuwei.shen@wmich.edu

© Springer International Publishing Switzerland 2015

R. Lee (ed.), *Computer and Information Science*, Studies in Computational
Intelligence 566, DOI 10.1007/978-3-319-10509-3_10

and error-prone. One of the important reasons is that developers should recall the details made during the software development process in order to build an argument linking the evidence and the corresponding claim(s). Then, a question has been raised: can an assurance case be generated when a specific software process has been employed?

In this paper, we take medical device software into account. In the medicals device industry, all medical devices must pass the FDA pre-market review before a new product can be deployed to the market because the FDA regulators are entitled to ensure that each new product is safe and reliable to the public. To better design a software system embedded into a medical device, some international standards such as ISO14791 [1] have been proposed. Recently, some guidance documents based on a specific medical device such as infusion pumps were released [2]. These standards and guidance documents aim to help medical device manufacturers to design medical device software according to recommendation in these documents. Consequently, the quality of medical device software can be improved.

However, how to conform to the regulation-requirements in these standards and guidance-has bothered many medical device manufacturers. In the medical device industry, many companies have already put in place their own cultures such as their own policies and processes to achieve the objectives of the regulation. They have found that it is a painful process to build a compelling assurance case which consists of information scattered over a large body of artifacts such as the hazards-analysis report and test and validation report, which they have produced before. Even worse, when constructing an assurance case, developers must spend more time to recall many design decisions made before. For instance, why a requirement has been decomposed into several sub-requirements and what criterion has been used at that time. Recall of these design decisions is always time consuming and error prone. Thus, the construction of an assurance case becomes one of persistent complains from medical device manufactures and stifle the motivation and creativity of manufacturers in building safer and more reliable medical devices.

In this paper, we strive for a new method that can automatically combine a specific software development process with the assurance information to reduce the burden of medical device manufacturers. We consider the goal structure notation (GSN) [3] as a backbone to build an assurance case. In order to integrate a specific development process employed by a medical device manufacturer, we employ the Rational Unified Process (RUP) [4] as a representative development process. We illustrate our approach with the case study of the Generic Insulin Infusion Pump (GIIP) [5], which we co-developed with FDA staff in the past years. In fact, in our previous work, the traceability established between different artifacts produced during the RUP process enable to track a system's design downstream to the implementation and upstream to the rationale. The traceability information leverages the understanding of how a system has been designed to satisfy the relevant safety requirement and thus expedite the regulator review process. Simultaneously, the traceability information becomes an integral part of constructing an assurance case.

The remainder of this paper is organized as follows: In Sect. 2, we introduce an assurance case and its GSN notation. In Sect. 3, we employ the Generic Insulin Infusion Pump case study to illustrate how to produce an assurance report when the Rational Unified Process is applied. Some related work is discussed in Sect. 4, and we draw a conclusion in Sect. 5.

2 Assurance Case and Its GSN Notation

An assurance case is important for safety critical systems in that it provides an argument from the developers about why a safety critical system can work well when it is in use. In the medical device industry, an assurance case of a medical device should demonstrate that the device does not harm a patient but also improve a patient's health. In order to assure the correctness of safety-critical software systems, a convincing assurance case should consist of the following elements:

• Claim. A claim is a statement that claims about some important property of a system such as safety and security.
• Argument. An argument is reasoning of why a claim can be supported. The reasoning can be done via the justification between claims and sub-claims or claims and evidence.
• Context. A context gives assumptions made about the whole assurance case.

A convincing and valid argument of why a system meets its assurance requirements is the heart of an assurance case. An assurance case should consist of extensive references to evidence used by the system. In general, an assurance case is a collection of claims, arguments, and evidence that are created to support the contention that a system will satisfy the particular requirements.

Certain claims can be directly supported by evidence, which usually refers to external documents collected by some systematic methods and procedures. In general, a structure of an assurance case is a tree-like structure with the top element as the root claim. Currently, there are two different notations to denote an assurance case, i.e. Goal Structuring Notation (GSN) [3] and Claims-Arguments-Evidence (CAE) Notation [6, 7].

In this paper, we adopt the GSN notation to represent an assurance case. In GSN, a rectangle node represents a claim, such as C0 ("All relevant hazards have been considered") in Fig. 1. A parallelogram node represents argument reasoning. In Fig. 1, the parallelogram A0 represents that the following argument: the decomposition of claim C0 is based on the categories recommended by the Guidance for Infusion Pump, i.e. "Total Product Life Cycle: Infusion Pump—Premarket Notification Submission" (abbreviated as Guidance in the rest of the paper) [2]. A rounded rectangle node denotes the relevant information used in an assurance case. For instance, C1 in Fig. 1 denotes the Guidance [2]. A diamond decorator node represents that information related to the node will be supported later. Finally, a circle node denotes evidence.

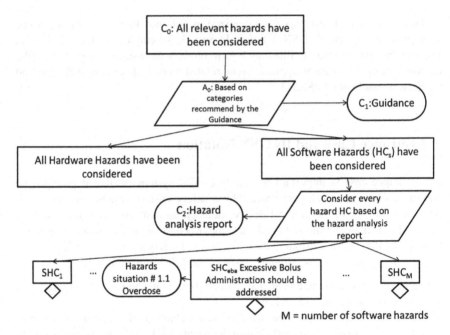

Fig. 1 *Top* level structure of an assurance case for GIIP

3 Development Process for GIIP

We apply the Rational Unified Process (RUP) to develop the Generic Insulin Infusion Pump case study due to the following several reasons. First, RUP is an iterative software development process created by Rational Software Cooperation, which was bought by IBM in 2003. It consists of various types of activities, each of which produces one or several different artifacts. RUP can be adapted to satisfy different needs during a software development process. Second, RUP has a good tool-based support. Various tools from IBM support different activities during software development. For example, RequisitePro [8] supports to produce artifacts at early stage of a software development process while Rational Software Architect (RSA) [9] can leverage the model design and software implementation. Because of the above reasons, RUP has gained a great popularity in industry.

In general, the RUP has four phases for a software development cycle. The key to the RUP is that a traditional software development process lies within all of the four phases, each of which has its own objective and milestone at the end. The milestone of each phase can be checked via the corresponding artifacts to validate whether the objective has been accomplished or not. These four phases are: Inception, Elaboration, Construction, and Transition. During the Inception phase, the primary objective is to scope the system adequately as a basis for validating initial costing and budgets. The primary objective of the Elaboration phase is to mitigate the key risk items; and at the same time, domain analysis is made and the architecture of the project gets its basic form. Consequently, some design artifacts such as a USE CSAE model should be available for the later phases. The primary objective of the Construction phase is to build the software system. The main activities should include the bulk of coding activities. As a result of the activity, software implementation as software artifacts should be done. Finally, during the Transition phase, the objective is "transit" the system from development into production, making it available to and understood by the end user. The main activities include to beta testing the system to validate it against the end users' expectations. The corresponding test and validation report should be produced.

The GIIP problem was proposed as a case study by FDA to study how a software system embedded into an insulin infusion pump can be designed and validated. In addition to the requirements related to an insulin infusion pump application, a software system designed for the GIIP application should also consider some specific standard or guidance documents in the medical device domain, such as "Total Product Life Cycle: Infusion Pump—Premarket Notification Submission" [2]. In this paper, we will demonstrate how to generate an assurance case when the RUP process has been employed.

First, an assurance case is necessary for each premarket submission from an infusion pump manufacturer. According to the Guidance, it says: "In making this demonstration of substantial equivalence for your infusion pump, FDA recommends that you submit your information through a framework known as an assurance case or assurance case report. An assurance case is a formal method for demonstrating the validity of a claim by providing a convincing argument together with supporting evidence." Furthermore, a claim, according to the Guidance, is "a statement about a property of the system". To make a claim, the Guidance identifies 8 different hazard categories. FDA further requires each submission to "clearly describe the method used to analyze the hazards and each hazardous event mitigation". Next, we illustrate how an assurance case can be generated during the RUP process. Due to the space limit, we cannot show all the artifacts produced during the RUP. We concentrate on the Inception/Construction Phase, which mainly produces the requirement document, feature description document, use case report, and a sequence diagram. As the starting point, we claim that all relevant hazards have been considered during the Construction Phase. Then, this claim can be decomposed into eight sub-claims according to the Guidance such as Hardware Hazards. The argument of this decomposition is shown by A0 and C1 in Fig. 1. In this paper, we only consider Software Hazards denoted as HCs. The Software Hazards can also be divided into some sub-subclaims and in this case we consider the "Excessive Bolus Administration" denoted as SHC_{eba}.

During the Construction Phase, first the safety requirement document should be produced. In this document, all safety requirements related to an infusion pump should be addressed. In the case of claim SHC_{eba}, the following argument is built: "Each hazard should be well addressed in a hazardous event mitigation way via safety requirements". In this argument, the phrase "well addressed" means as follows: the disjoint of relevant subclaim/sub-arguement/evidence should be complete (exhaustive) while the conjoint of the subclaim/sub-arguement/evidence should be

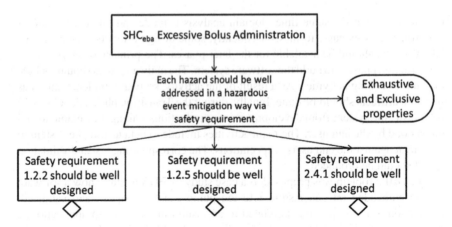

Fig. 2 Structure between SHCeba and safety requirements

empty (exclusive). More specifically, during the Inception Phase, a design decision about safety requirements should be made as follows: all safety requirements should satisfy the exhaustive and exclusive properties. If there is only one safety requirement produced:

- 1.2.2 The pump shall allow the user to set at least two basal profiles at the same time, and require the user to activate no more than one profile at any single point in time.

then the above argument is not complete since how the hazard SHC_{eba} can be mitigated when a user set only one basal profile is not addressed. So, when developers apply the RUP to produce the safety requirements documents, the flaw in the assurance case is observed. Figure 2 shows the structure between hazard SHC_{eba} and the related safety requirements.

Assume that the other safety requirement related to when a user sets one basal profile is considered continue to apply the RUP to design features document. A features document lists all the features that an infusion pump system should achieve. Similar to the previous argument structure, we should build an argument that "Each safety requirement should be well addressed via system features". In this case, we assume that the following features are produced:

- R3113-2: The component shall be able to manipulate the Basal Profiles record in the following ways:

  - Add a new profile to the record if doing so will not exceed the record's capacity (see Requirement R3116 for more detail).

- R3114: The component shall not accept any invalid basal profile that the user programs into the Basal Profiles record. A valid basal profile includes one or more segments, each of which is defined as a pair (effective period, basal rate), where the basal rate shall range from 0.05 Unit/h to x Unit/h and the effective period is defined by its starting time (of day) and ending time (of day). The ending time of a basal segment shall be no earlier than its starting time. In a valid basal profile, the effective periods of two distinct segments shall not overlap with each other, and effective periods of all segments shall cover 24 h of day.

- R3115: If the Basal Profiles record is not empty, the component shall allow the user to activate (via IUID) any profile stored in the record, i.e. scheduling basal delivery according to this profile. If the profile to be activated is not currently active, the component shall deactivate the currently active one first and then activate this profile. The activation of basal profiles has to be confirmed by the user before it can take effect.

- R3116: When the user selects (via IUID) to program a new basal profile, the component shall check if the Basal Profiles record has reached its storage limit. If so, the component shall instruct IUID to prompt the user to either quit programming or select an existing profile to override. Otherwise, the component shall acquire the newly programmed profile from IUID and check its validity. If the new profile is valid, the component shall add it into the Basal Profiles record without affecting other profiles in the record or selection of the active basal profile. If the new profile is not valid, the component shall reject this profile and inform IUID of the rejection.

In the case of claim safety requirement 1.2.2 is well designed, two arguments are built based on the fact that each safety requirement is mapped to a set of features that satisfy the exhaustive and exclusive properties and each safety requirement should be connected with related features. For the first argument, the exhaustive and exclusive properties mean that each feature considers one scenario for a safety requirement. For example, safety 1.2.2 can be decomposed into the following scenarios that should be addressed by the features: (1) A function that allows the user to add new profiles, (2) The definition of a valid profile that can be accepted by the system, (3) A function that allows the user to activate profile, and (4) the constraint of activating a profile. In this example, R3113-2 considers adding a new profile, which is further explained by R3116. These two features are related to the scenario 1 of safety 1.2.2. The scenario 2 is addressed by R3114 which defines a valid structure of a profile. When a basal profile is added, R3115 shows the feature of setting an added basal profile to be an active basal profile and is related to scenarios 3 and 4 of safety 1.2.2. From these descriptions, the design decision is based on that these features should satisfy the exclusive and exhaustive properties. Namely, any two of these features have no overlapping, and safety requirement 1.2.2 is completely addressed by these features. For the second argument, the traceability links between safety requirements and system features are considered as evidence to support this argument. Figure 3 shows the arguments and sub-claims decomposed from safety requirement 1.2.2. Figure 4 shows the traceability matrix between safety requirements and system features.

Next, we establish an argument via a design decision made about features via some artifacts produced during the next activity of the RUP as shown in Fig. 5. In this case, we enter the Construction Phase in the RUP which mainly produces some design artifacts. Some important artifacts to be produced include use case reports,

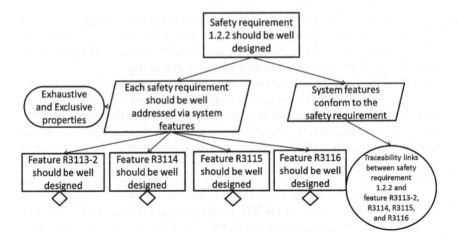

Fig. 3 Structure between safety requirement 1.2.2 and features

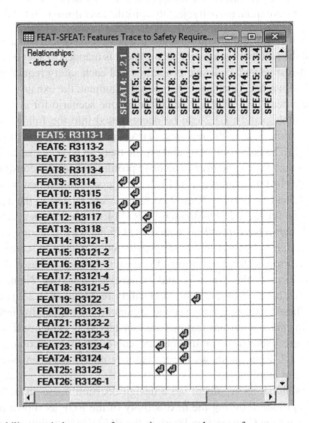

Fig. 4 Traceability matrix between safety requirements and system features

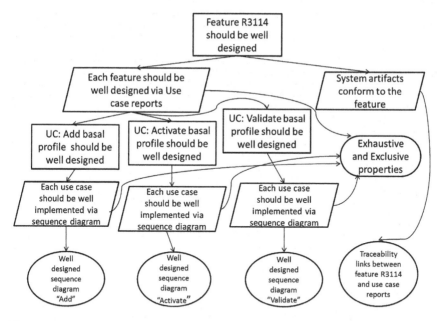

Fig. 5 Structure of feature R3114

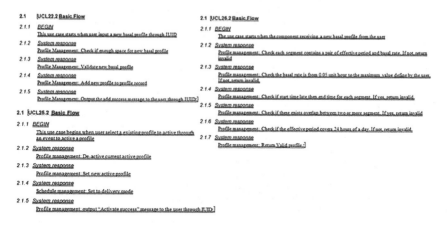

Fig. 6 Use case reports related to feature R3114

and sequence models. Likewise, we can continue to consider the similar arguments, each feature should be well designed via a use case report and each feature should be connected with related system artifacts. Here we use Feature R3114 as an example. Figure 6 shows three use case reports that targets on the three different scenarios related to R3114 and Fig. 7 shows the traceability matrix between features and use case reports.

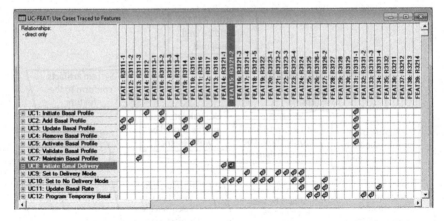

Fig. 7 Traceability matrix between system features and use case reports

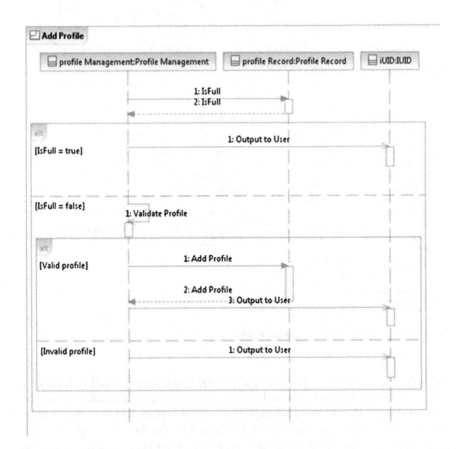

Fig. 8 Sequence diagram for add profile

Last, due to space, we assume that the RUP process stops at the second argument, i.e. each use case is designed. In the real development process, we continue to develop the system in which a design class model and a final C++ implementation are produced. Under this assumption, these sequence diagrams serve as evidence of the whole argument hierarchy. The logical connection between a high-level hazard and the final sequence diagram reflects how each hazard has been analyzed and mitigated. We show a sequence diagram related to Use Case "Add basal profile" in Fig. 8. Here, it is worth discussing what "each use case should be well designed" means. We follow the definition of exhaustive and exclusive properties to design sequence diagrams from use case models. In order to comply with the exclusive property, every use case is implemented by different sequence diagram. Each sequence diagram describes different system interactions and no overlapping between any two sequence diagrams. To fulfill the exhaustive property, every basic flow and alternative flows of a use case can be mapped to different message in the sequence diagram. Every lifeline designed in the sequence diagram is already defined in the use case. That is, all of the information in the use case is completely captured and implemented by the sequence diagram.

4 Related Work

Construction of an assurance case has been a hot topic since the software-intensive safety-critical systems are becoming popular and increasing complex. How to build a compelling argument draws more attention in safety-critical industry. Many challenges have remained in the assurance case community [10]. Various techniques have been proposed to address the problems and difficulties in the construction of an assurance case.

Hawkins et al. presented a new approach that incorporates a confidence argument in a safety argument [11]. Traditionally, a safety argument includes the confidence about this argument and the separation of a confidence argument can make both arguments clarity of purpose and helps to avoid some redundancy in arguments and evidence.

Jee et al. discussed the construction of an assurance case for the pace-maker software via a model-driven development approach [12], which is similar to ours. However, their approach emphasizes on the later stage of a software development such as a timed automata model as a design model and C code as implementation language. The approach considers the application of the results from a model checker called UPPAAL and measurement based on timing analysis as evidence.

Attwood et al. proposed to apply a linguistic model of understanding to identify mismatches and provide guidance on composition and integration when constructing an assurance case. Dominguez et al. presented an experience in developing an assurance case for a rebreather system via the Goal Structuring Notation. Ray et al. demonstrate an approach for safety assurance case argumentation based on the Generic Patient Analgesic Pump (GPCA). As a governing agency in medical device industry, a document recently released by FDA recommends the submission of an assurance case as part of pre-market submission.

Rushby observed that an assurance/safety case is an argument that helps increase confidence in the soundness of a given case. One of the most important ways to check an argument is formal logic. Thus Rushby proposed a method to formalize safety cases and one important ramification of this work is the use of automated tools to check the logical soundness of a safety case [13]. Another advantage of formalization is the development of metamodel for various tactics of argument.

The automotive industry is another specific domain requiring the construction of a safety case. For example, the automotive standard ISO26262 [14] requires the development of a safety case for electrical and/or electronic (E/E) systems whose malfunction has the potential to lead to an unreasonable level of risk. Many researchers have developed various strategies to design a safety case for an automotive software system. Birch et al. investigated the main argument structures of a safety case and the relationships among these structures when assessing functional safety in that ISO26262 does not specify how a safety argument should be evaluated in the functional safety assessment process [15]. Birch et al. emphasized on the product-based safety rational when constructing a safety argument.

Westman et al. demonstrated that the contract theory can be employed to construct safety requirements in ISO26262 [16]. Contracts are used to separate the responsibility of a system from its environment by imposing safety requirements on the environment as assumptions. To check an automotive software system against ISO 26262, contract theory provides the verification of consistency and completeness on the safety requirements.

Stürmer et al. proposed a novel approach to study whether an automotive software system is compliant with ISO26262 via reviewing software models [17]. Since model-based development has gain the popularity in the automotive industry, the early detection of model artifacts that violate the safety requirements in ISO26262 can greatly improve the quality of an automotive software system. Stürmer et al. combined an automated and manual review to detect any violation of ISO26262.

5 Conclusion and Future Work

Developing a software system that, software engineers can guarantee, satisfies the regulatory requirements is one of the most challenging issues facing the software engineering community. In this paper, we propose a novel approach which integrates the construction of an assurance case into a software development process. While we only consider the Rational Unified Process, our approach can be applied to all other software development processes employed by medical device manufacturers. We aim to save the time and labor to generate a convincing and compelling argument for a system and our approach cannot only be used by some regulatory bodies but also improve manufacturers' capability to understand the quality of a system they have designed.

While our approach is still in the preliminary stage, we think the approach paves the way to leverage the capability of regulatory review and even software certification via an assurance case in an automatic way. With the recent progress made in the Model Driven Engineering (MDE) community, we think some important techniques advanced in the MDE community can be adapted to regulatory review and software certification. In both regulatory review and software certification, one important issue is to investigate whether a system has achieved the claimed requirements. In fact, an assurance case has been widely proposed to establish an argument in a logically consistent fashion.

We have noticed that a metamodel for a structured assurance case, called SACM [18], has been proposed by the Object Management Group while GSN and CAE have been used as popular notations in the assurance case community. Obviously, the assurance case community does not lack of the notation to represent an assurance case. The introduction of SACM is in fact consistent with the latest development fashion in MDE, aiming to bridge the gap between the problem domain and the implementation domain.

On the other hand, a UML profile mechanism has been widely employed to model different artifacts produced during software development process. Also, a UML profile can model a development process. For instance, a UML profile for business modeling proposed by IBM aims at the application of UML notation to represent the artifacts produced during BPMN [19]. In fact, many industry companies have their own metamodel capturing the specific development process used in their company. Based on this fact, we think the application of the metamodels that represent a development process and an assurance case structure respectively can facilitate the regulatory review and software certification from the following two aspects.

First, inspired by the forward engineering features in MDE, we take the advantage of the two different metamodels that can help to generate an assurance case. One important feature of a metamodel is to leverage the ability of model transformation. Thus, using the model transformation techniques we can produce an assurance case from artifacts produced during a development process. In this case, developers can record all design decision made to produce various artifacts during a development process. Consequently, there is no additional time and effort to build an assurance case separately.

Second, the popular technique in MDE to retrieve a design model from an implementation motivates us to retrieve an assurance case from the produced artifacts. This so-called reverse engineering is quite useful when a system has been designed before some standards and guidance documents are proposed. Obviously, the metamodels can help to dive into text in the related documents that can be possibly established an argument structure. Next, thanks to the latest development in information retrieval, we can consider to apply some techniques such as vector space model to retrieve the relevant information to recover an assurance case.

An assurance case provides a powerful method for medical device manufacturers to convince some regulation agencies such as FDA that their system is compliant with safety requirements under some guidance documents or standards. Our future work based on the approach will concentrate on the application of the MDE techniques. Thus, the generation of an assurance case can be done in an automatic and systematic approach so we should finally leverage the capability of regulatory review and software certification.

References

1. Medical devices—Application of risk management to medical devices, ISO 14971
2. US Food and Drug Administration, Guidance for Industry and FDA Staff-Total Product Life Cycle: Infusion Pump- Premarket Notification[510 (k)] Submissions. April 2010
3. Kelly, T., Weaver, R.: The Goal Structuring Notation—A Safety Argument Notation, in dependable systems and networks 2004 workshop on assurance cases (2004)
4. Kruchten, P.: The Rational Unified Process: An Introduction. Addison-Wesley Professional, Amsterdam (2003)
5. FDA, Generic Insulin Infusion Pump Functional Specifications (2011)
6. Adelard. The Adelard Safety Case Editor—ASCE. http://adelard.co.uk/software/asce/ (2003)
7. Bishop, P.G., Bloomfield, R.E.: The SHIP Safety Case Approach, in Safe Comp 95, pp. 437–451. Springer, London (1995)
8. Zielczynski, P.: Requirements Management Using IBM Rational RequisitePro. IBM Press, Upper Saddle River (2008)
9. Leroux, D., Nally, M., Hussey, K.: Rational software architect: a tool for domain-specific modeling. IBM Syst. J. **45**(3), 555–568 (2006)
10. Langari, A., Maibaum, T.: Safety Cases: A Review of Challenges (2013)
11. Hawkins, R., Kelly, T., Knight, J., Graydon, P.: A New Approach to Create Clear Safety Arguments, In Nineteenth Safety-Critical Systems Symposium. Southampton, UK (2011)
12. Jee, E., Lee, I., Sokolsky, O.: Assurance Cases in Model-Driven Development of the Pacemaker Software, LNCS 6416 (2010)
13. Rushby, J.: Formalization in Safety Cases. In Eighteenth Safety-Critical Systems Symposium, pp. 3–17 (2010)
14. CD ISO, Road vehicles-Functional safety, International Standard ISO/FDIS, vol. 26262 (2011)
15. Birch, J.: Safety cases and their role in ISO 26262 functional safety assessment. In Computer Safety, Reliability, and Security, pp. 154–165. Springer (2013)
16. Westman, J., Nyberg, M., Törngren, M.: Structuring safety requirements in ISO 26262 using contract theory. In Computer Safety, Reliability, and Security. pp. 166–177, Springer (2013)
17. Stürmer, I., Salecker, E., Pohlheim, H.: Reviewing software models in compliance with ISO 26262. In Computer Safety, Reliability, and Security. pp. 258–267, Springer (2012)
18. OMG. Structured Assurance Case Metamodel (SACM)—Version 1.0. http://www.omg.org/spec/SACM/
19. Johnston, S.: Rational UML Profile for business modeling, IBM Developer Works. http://www.ibm.com/developerworks/rational/library/5167.html, (2004)

A Survey on the Categories of Service Value/Quality/Satisfactory Factors

Yucong Duan, Nanjangud C. Narendra, Bo Hu, Donghong Li,
Wenlong Feng, Wencai Du and Junxing Lu

Abstract Service modeling relies on the many factors as indicators for measurement of business value, service quality and user satisfaction. However standardization on related factors is missing in current literature. In this paper, we work towards the solution of this problem. Firstly we collect more than 200 related factors from literature review. Secondly we try to provide a classification framework through a constructive process at conceptual level. Then we use the constructive process to classify the factors into more than 20 higher level categories with explanation on the process.

Keywords Service value · Service quality · Satisfactory · Classification

Y. Duan (✉) · W. Feng · W. Du · J. Lu
College of Information Science and Technology, Hainan University, Haikou, China
e-mail: duanyucong@hotmail.com

W. Feng
e-mail: fwlfwl@163.com

W. Du
e-mail: wencai@hainu.edu.cn

J. Lu
e-mail: lujx1994@gmail.com

N.C. Narendra
Cognizant Technology Solutions, Bangalore, India
e-mail: ncnaren@gmail.com

B. Hu
Kingdee International Software Group China, Hong Kong, China
e-mail: bob_hu@kingdee.com

D. Li
School of Statistics and Mathematics, Central University of Finance and Economics,
Beijing, China
e-mail: lidh97@cufe.edu.cn

© Springer International Publishing Switzerland 2015   141
R. Lee (ed.), *Computer and Information Science*, Studies in Computational
Intelligence 566, DOI 10.1007/978-3-319-10509-3_11

Y. Duan et al.

1 Introduction

Service measurement is of key importance to the successful adoption of various service technologies.

This is because service applications must provide clear indicators for profit calculation before any investors will be willing to invest in them. From the software management perspective, a service system can be defined as successful only when the end users are satisfied, and this must be measurable. For service system to be maintained quality factors such as QoS need to be employed as technical requirements. In the service market, service quality factors will be evaluated in a comprehensive manner before candidate services are selected for a service composition.

Ideally a standardization of service factors which coherently cover all aspects in a coherent manner is expected to aid the widespread use of service technologies. To the best of our knowledge, there is no dedicated literature for this purpose. In this paper, we work towards this direction starting from an empirical collection of the factors and proceed towards creating an approach to classify the factors.

The rest of the paper is organized as follows: Sect. 2 shows the empirically collected factors. Section 3 presents our classification approach and the construction steps. It also shows the application of our approach. Section 4 concludes the work with future directions.

2 Emprical Collection of Factors

Table 1 shows a collection of more than 200 service value/quality/satisfactory factors from literature review. We did this by going through the main literatures in both IT and economics research.

Empirically the main categories are collected as follows:

{Functional, Implementation, emotional, epistemic, social, management, reshaping, quality, context, security, supportive, computation, risk, input, monitor and control, integration, composition, legislation, compliance, financing, alter perspectives, tradeoff}

3 Organization of the Categories

Figure 1 shows the organization of the categories which actually gives a certain degree of formal meaning (although not complete) to these natural language terms by relating them together in a hierarchy. The organization is explained with the subsequent decomposition steps.

Table 1 Collection of service value/quality/satisfactory factors

Category	Sub category
Functional	Practical, correct, state, event, activity, exchange, usage, operation
Implement (roles)	Adaptation, order, creation, deployment, activate, terminate, charge, report, purchase, response, sell, marketing, transfer, design, interface, decision, development
Emotional	Dependent, partnership, friendliness, cooperativeness, confidence, brand, trust, satisfaction, anxiety, surprise, shame, psychological, imagination, inspiration
Epistemic	**Description**, define, knowledge, organization, service type, data classification, **learn, understanding, reuse, metamodel, cognitive, difficult, identification, prioritize, innovation**
Social	Reputation, trust, culture, group, peer pressure, community, fashion
Management	Test, validation, disposition, withdraw, expiration, report, alarm, derivation, collection, reproduction, evaluation,
Reshaping	Mature, habit, feedback, reengineering, repurchase, recommendation to others, persuasion
Quality	Responsiveness, completeness, compatibility, consistency, up-to-date, readability, availability, precision, extensibility, granularity, accuracy, effective, efficient, capability, aesthetics
Context	Restriction, location, domain, language, format, technique, feature, conditional, member ship, customer loyalty
Security	Privacy, data integrity, encryption
Supportive	Remedy, compress, revision, logging, renew, Archiving, backup, Recovery, extension, optimization, replace, history, environment, elicitation, statistics, insurance, infrastructure
Computation	Logical, measure, priority, classification, precondition, post condition, assessment, *sacrifice, benefit*
Risk	Uncertainty, indeterministic, possibility, change management
Input	Price model, Questionnaire, cost, requirement, supply mode, right, ownership, expectation, incentive, intention, coverage, time, size, name, organization, contract, quantity
Monitor and Control	Behavior, performance, governance, reconciliation, communication, lifecycle, relevance analysis, response time, update frequency, simulation, checking
Integration	Information/ data Integration, resource planning
Composition	Functionality(de) composition, **business process**, *subscription/publish, transaction, networked, transformation, bundling*
Legislation	Law, policy, **subcontract,** liability, enforcement, copyright, license, intellectual right, jurisdiction, indemnity
Compliance	Evaluation practice, SLA, obligation, responsibility, credit, standard, protocol, agreement, usage permission, warranty, non-commercial, attribution, commit
Financing	*Payment/cost/charge management, audition, tax, accounting, value, added value, created value, cost model*
Alter	*Service barter, advertisement, services oriented network*
Tradeoff (roles)	*Long term versus short term, stakeholders, roles, sustainability, game theory, perspectives, reciprocity, conflicts, promotion, strategy*

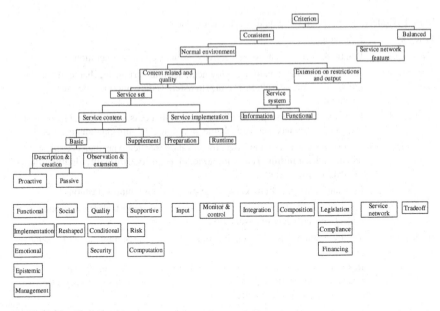

Fig. 1 Evaluation criteria

3.1 Classification According to "Consistent" Versus "Balanced"

Source: this classification is targeted on the identification of whether a service model considers competitive situation.

Description:

Consistent: it means that a service model can be placed into at least one subcategory which belongs to the following set of considered subcategories:

{Functional, Implement, emotional, epistemic, social, management, reshaping, quality, context, security, supportive, computation, risk, input, monitor and control, integration, composition, legislation, compliance, financing, alter perspectives}

Balanced:it means that a service model covers at least one considered subcategory of tradeoffs among no less than two parties, such as:

{roles, stakeholders, goals, interests}

Considered set of subcategories of balanced:

{tradeoff}. It refers to tradeoffs among long term versus short term goals, tradeoff among interests of roles/stakeholders, competition among companies.

Usage of this classification: it can used to indicate the coverage of the content of a service model by marking one or more of elements under {consistent, balanced}.

3.2 Under "Consistent" Category: Classification According to "Normal Environment" Versus "(Plus) Service Network"

Source: this classification is targeted on the identification of whether a service model is running in an environment with all the interaction interfaces which are provided in the form of service interfaces which we call service network infrastructure, or in a normal environment.

Description:

Service network: the infrastructure for a service model to run is built on services by itself. So all the interfaces which are provided to a running service model are service interfaces. In this environment, a service model is an instance of SaaS (software as a service) in cloud computing.

Normal environment: it means that a service model is not in a service network for this evaluation approach.

Considered set of subcategories of service network:

{service network}

Usage of this classification: it can used to indicate the coverage of the content of a service model by marking Yes/No on {service network}. If service network is marked as Yes, the subcategories of "normal environment" might still be applied to evaluate a service model which is classified under "service network" because that most of those subcategories which are taken as evaluation criteria are used to measure the coverage of a service model or its parts and are independent from the confirmation of Yes of "service network" of a service model.

Notice: a "service network" is an extension of a "normal environment". "Normal environment" is not an explicit option for selection.

3.3 Under "Normal Environment" Category: Classification According to "Content Related and Quality" Versus "Extension on Restrictions and Output"

Source: this classification is targeted on whether a service model covers solely the content and quality of a service system or even more by covering the contractual restrictions and economical activities based on the output data of services from business perspective.

Description:

"Content related quality": it refers to that a service model covers the information related directly to the construction, implementation and quality evaluation related issue.

"Extension on restrictions and output": in general, it refers to the issues which are independent from a concrete service system but can be adopted for implementation of a service system. Restriction refers to the external restrictions on operations on service systems which are irrelevant to the concrete implementation, construction and quality of a specific service system. Extension on output refers to the economic and financing considerations and activities which are based on the output data of a service system.

Considered set of subcategories of "content related and quality": {Functional, Implement, emotional, epistemic, social, management, reshaping, quality, context, security, supportive, computation, risk, input, monitor and control, integration, composition}

Considered set of subcategories of "extension on restrictions and output":

{legislation, compliance, financing}

Usage of this classification: it can used to indicate the coverage of the content of a service model by marking Yes/No on {extension on restrictions and output}. If "extension on restrictions and output" is marked as Yes, all subcategories of "content related and quality" might still be applied to evaluate the service model.

Please note: an "extension on restrictions and output" is an extension on solely "content related and quality".

3.4 Under "Content Related and Quality" Category: Classification According to "Service Set" Versus "Service System"

Source: this classification is targeted to refine the evaluation of a service model as objects of "service set" and "service system".

Description:

"Service set": it refers to those evaluations that focus on individual services or groups of services which will consider the integration of organized of services in the form a service system.

"Service system": it refers to those evaluations that focus on the structure and performance of organized of services in the form a service system instead of individual services or groups of services.

Considered set of subcategories of "service set":

{Functional, Implement, emotional, epistemic, social, management, reshaping, quality, context, security, supportive, computation, risk, input, monitor and control}

Considered set of subcategories of "service system":

{Integration, composition}

Usage of this classification: it can used to refine the evaluation on a service system by isolating features which solely belong to a service system instead of set of services which compose a service system. It can used to indicate the coverage of the content of a service model by marking Yes/No on {service system}. If service system is marked as Yes, the subcategories of "service set" can still be applied to evaluate a service model since under the category of service system only issues which are not covered by the category of service set are considered.

Execution rules:

- When considered for this classification, first evaluate under the category of "service set".
- After evaluation under "service set", consider evaluation under category of "service system".

3.5 Under "Service System" Category: Classification According to "Information" Versus "Functional"

Source: this classification is targeted on refinement of characteristics of service system instead of service sets under the perspectives of information versus functional.
Description:
Information: it refers to the integration of information and data in a service system.
Functional:it refers to the composition of functions of individual services to form the functions at the service system level.
Usage of this classification: it can used to indicate the coverage of the content of a service model by marking Yes/No on {information, functional}. The marking on both of these categories are fully compatible since that they are independent from each other.
Considered set of subcategories of "information":
{integration}. It refers to the integration of information and data inside a service system during runtime.
Considered set of subcategories of "functional":
{functional}. It refers to the composition of interactions of services inside a service system during runtime.

3.6 Under "Service Set" Category: Classification According to "Service Content" Versus "Service Implementation"

Source: this classification is targeted on distinguishing between characteristics of service content and implementation of services.
Description:
Service content: it refers to the issues which are related to the creation, observation of individual services which compose a service model.
Service implementation: it refers to the issues which are related to the implementation of individual services which compose a service model.
Usage of this classification: it can used to indicate the coverage of the content of a service model by marking Yes/No on considered subcategories under {service content, service implementation}.
Considered set of subcategories of "service content":

{Functional, Implement, emotional, epistemic, social, management, reshaping, quality, context, security, supportive, computation, risk}

Considered set of subcategories of "service implementation":

{Input, monitor and control}. Input refers to the input preparation for individual services from users and other service systems. Monitor and control refers to the monitor and control of individual services at runtime.

3.7 Under "Service Content" Category: Classification According to "Basic" Versus "Supplement"

Source: this classification is targeted on refinement on service content as basic issues and supplement issues.

Description:

Basic: it refers to the issues which are related to the description, creation, observation and extensions of individual services which compose a service model.

Supplement: it refers to the issues which are considered but not under the category of "basic".

Usage of this classification: it can used to indicate the coverage of the content of a service model by marking Yes/No on considered subcategories under {basic, supplement}.

Considered set of subcategories of "basic":

{Functional, Implement, emotional, epistemic, social, management, reshaping, quality, context, security}

Considered set of subcategories of "supplement":

{Supportive, computation, risk}.

3.8 Under "Basic" Category: Classification According to "Description & Creation" Versus "Observation & Extension"

Source: this classification is targeted on refinement on basic issues of a service as related to the content description, service creation and supplement issues.

Description:

Description & creation: it refers to the issues which are related to the description of the content of a service, and management of the creation process.

Observation & extension: it refers to the issues which are related to the quality of service, conditional issues of service content such as location, available time, data quality and security.

Usage of this classification: it can used to indicate the coverage of the content of a service model by marking Yes/No on considered subcategories under {description & creation, observation & extension}.

Considered set of subcategories of "description & creation":
{Functional, Implement, emotional, epistemic, social, management, reshaping}
Considered set of subcategories of "observation & extension":
{Quality, context, security}.

3.9 Under "Description & Creation" Category: Classification According to "Proactive" Versus "Passive"

Source: this classification is targeted on refinement on the category of "description & creation" of a service as proactive or passive.

Description:

Proactive: it refers to the issues which are related to active description of the functionality of a service and implementation of creation of individual in a possibly managed process.

Proactive: it refers to the issues which are related to passive content and activities on a service. Examples include the social impact on a service and reshaping activities on service usage.

Usage of this classification: it can used to indicate the coverage of the content of a service model by marking Yes/No on considered subcategories under {proactive, passive}. The marking on both of these categories are fully compatible since that they are independent from each other in the context of this approach.

Considered the set of subcategories of "proactive":
{Functional, Implement, emotional, epistemic, management}
Considered set of subcategories of "passive":
{Social, reshaping}.

Meaning of specific criteria items:

For all the criteria items under the subcategories in this approach, their meanings are restricted by the meaning of the subcategories.

Figure 1 shows an initial application of the approach to classify and evaluate the current literature. The information gained from this survey will be useful to evaluate not only individual work in terms of coverage but also can be used to identify the important factors to which we need to be paid more attention.

Table 2 A survey on the usage of service value/quality/satisfactory factors

category	Functional	Implement	Management	Emotional	Epistemic	Social	Reshaping	Quality	Context	Security	Supportive	Computation	Risk	Input	Monitor and Control	Integration	Composition	Legist Ration	Compliance	Financing	Alter	Tradeoff
Batini et al. [1]	V	V	V					V	V		V	V		V	V	V	V	V	V	V		V
Andersson et al. [2]	V	V	V					V	V				V	V	V	V	V	V	V	V		V
Ruiz et al. [3]			V	V	V	V	V	V	V		V	V		V	V		V			V		V
Oh [4]	V		V	V		V	V			V	V			V		V				V		V
Lin et al. [5]		V		V		V	V	V	V	V	V	V		V	V				V	V		V
Sheth et al. [6]	V	V		V	V	V			V		V			V					V			
Zeithaml [7]	V	V	V	V	V		V	V	V		V			V	V					V		V
Allahbakhsh et al. [8]	V	V	V		V			V			V			V		V						
KSRI [9]	V	V	V	V		V		V	V			V	V	V			V		V	V		V
Andrikopoulos [10]		V	V					V	V		V			V		V			V	V		V
E3value [11, 12]	V	V	V	V	V	V		V	V	V	V	V		V	V	V	V	V	V	V	V	V
Manzoor [13]	V	V	V			V		V						V			V	V				
Baida [14]	V	V	V	V							V			V			V			V		
Kinderen and Gordijn [15]	V	V	V	V							V	V		V			V			V		
Gordijn [16]	V	V	V	V	V			V			V			V	V		V			V		V
Weigand et al. [17]	V	V	V	R	R	R		V			V	V		R	V	V	V		R	R		R
Edirisuriya and Zdravkovic [18]	V	V		V				V		V	V			V		V	V			V		V
Allee [19]	V	V	V	V	V			V	V		V			V		V	V			V		V
McCloughan and Lyons [20]	V							V	V	V	V	V		V								V
Caswell et al. [21]	V	V	V	V	V	V		V			V	V	V	V	V	V	V	V		V	V	V
Sum (occurrences)	17	19	15	14	11	8	4	16	13	3	13	15	4	15	15	7	17	6	8	17	2	16

V confirms the usage of related terms
R means loosely related

4 Conclusion and Future Work

This paper is just a first step of our ongoing work towards providing a measure framework and an approach to create it. We will continue to explore with more literature and refine the work as per our continuing evaluation. In particular, we will emphasize quantitative evaluation of these factors, with a view towards measuring service systems more rigorously. We hope that it will be beneficial to stakeholders for evaluating service systems and modeling service projects.

Acknowledgments This paper was supported in part by CNSF grant 61363007 and by HNU Research program grant KYQD1242 and HDSF201310.

References

1. Batini, C., Viscusi, G.: Planning egovernment information systems: methods and experiences from the eG4M framework. DG.O, pp. 371–372. ACM, New York (2011)
2. Andersson, B., Bergholtz, M., Edirisuriya, A., Ilayperuma, T., Johannesson, P., Gordijn, J., Grégoire, B., Schmitt, M., Dubois, E., Abels, S., Hahn, A., Wangler, B., Weigand, H.: Towards a Reference Ontology for Business Models ER, pp. 482–496. Springer, Berlin (2006)
3. Martín, D.R., Gremler, D.D., Washburn, J.H., Carrión, G.C.: Service Value Revisited: Specifying a Higher-order, Formative Measure. Journal of Business Research. JBR. (2008)
4. Oh, H.: Service quality, customer satisfaction and customer value: a holistic perspective. Int. J. Hospitality Manag. **18**(1), 67–82 (1999)
5. Lin, C.-H., Sher, P.J., Shih, H.-Y.: Past progress and future directions in conceptualizing customer perceived value. Int. J. Serv. Ind. Manage. **16**(4), 318–336 (2005)
6. Sheth, J.N., Newman, B.I., Gross, B.L.: Why we buy what we buy: A theory of consumption values. Journal of Business Research **22**(2), 159–170 (1991)
7. Zeithaml, V.A.: Consumer perceptions of price, quality, and value: A means-end model and synthesis of evidence. J. Marketing **52**(3), 2 (1988)
8. Allahbakhsh, M., Benatallah, B., Ignjatovic, A., Nezhad, H.R.M., Bertino, E., Dustdar, S.: Quality Control in Crowdsourcing Systems: Issues and Directions. IEEE Internet Comput. **17**(2), 76–81 (2013)
9. KSRI (Satzger et al).: Knowledge Intensive Services Procurement Strategy,KIT, http://www.ksri.kit.edu/downloads/Knowledge_Intensive_Services_Procurement_Strategy._KSRI_Research_Report.pdf (2009)
10. Andrikopoulos, V.: A Theory and Model for the Evolution of Software Services, CentER Dissertation Series. Tilburg University Press, Tilburg (2010)
11. Gordijn, J., Razo-Zapata, I.S., De Leenheer, P., Wieringa, R.: Challenges in Service Value Network Composition. pp. 91–100. PoEM (2012)
12. Gordijn, J., Yu, E.S.K., van der Raadt, b: E-service design using i* and e3value modeling. IEEE Softw. **23**(3), 26–33 (2006)
13. Manzoor, A., Truong, H.L., Dustdar, S.: On the Evaluation of Quality of Context. EuroSSC, pp. 140–153. Springer, Heidelberg (2008)
14. Baida. Z.: Software-aided Service Bundling—Intelligent Methods & Tools for Graphical Service Modeling. Ph.D. thesis, VrijeUniversiteit, Amsterdam, NL (2006)
15. de Kinderen, S., Gordijn, J.: e3service—a structured methodology for generating needs-driven it-service bundles in a networked enterprise. In: Symposium on Applied Computing (SAC'08), ACM, Fortaleza, Brazil (March 2008)

16. Gordijn, J., de Kinderen, S., Wieringa, R.: Value-driven Service Matching. pp. 67–70. RE (2008)
17. Weigand, H., Johannesson, P., Andersson, B., Bergholtz, M.: Value-Based Service Modeling and Design: Toward a Unified View of Services. pp. 410–424, CAiSE (2009)
18. Edirisuriya, A., Zdravkovic. J.: Aligning Goal and Value Models for Information System Design. pp. 126–140, MCETECH (2009)
19. Allee, V.: A value network approach for modelling and measuring intangible. In: Proceedings Transparent Enterprise (2002)
20. McCloughan, P., Lyons, S.: Accounting for ARPU: New evidence from international panel data. Telecommun. Policy $30(10)$, 521–532 (2006)
21. Caswell, N., Nikolaou, C., Sairamesh, J., Bitsaki, M., Koutras, G. D., Iacovidis, G.: Estimating Value in Service Systems—A theory and an example. IBM Systems Journal, vol 47, no 1. (2008)

Effective Domain Modeling for Mobile Business AHMS (Adaptive Human Management Systems) Requirements

Haeng-Kon Kim and Roger Y. Lee

Abstract Software development projects tend to grow larger and more time consuming over time. Many companies have turned to software generation techniques to save time and costs. Software generation techniques take information from one area of the application, and make intelligent decisions to automatically generate a different area. Considerable achievements have been made in the areas of object-relational mappers to generate business objects from their relational database equivalents, and vice versa. There are also many products that can generate business objects and databases using the domain model of the application. Domain engineering is the foundation for emerging "product line" software development approaches and affects the maintainability, understandability, usability, and reusability characteristics of family of similar systems [1]. In this paper, we suggest a method that systematically defines, analyzes and designs a domain to enhance reusability effectively in Mobile Business Domain Modeling (MBDM) in Adaptive Human Management Systems (AHMS) requirements phase. For this, we extract information objectively that can be reused in a domain from the requirement analysis phase. We sustain and refine the information, and match them to artifacts of each phase in domain engineering. Through this method, reusable domain components and malleable domain architecture can be produced. In addition, we show the practical applicability and features of our approach.

Keywords Mobile business domain engineering · Mobile domain analysis · Domain reuse · Component-based software development · Business domain model · Mobile business domain architecture

H.-K. Kim (✉)
School of Information Technology, Catholic University of Daegu, Hayang, South Korea
e-mail: hangkon@cu.ac.kr

R.Y. Lee
Department of Computer Science, Central Michigan University, Michigan, USA
e-mail: lee@cps.cmich.edu

© Springer International Publishing Switzerland 2015
R. Lee (ed.), *Computer and Information Science*, Studies in Computational
Intelligence 566, DOI 10.1007/978-3-319-10509-3_12

1 Introduction

Domain engineering supports application engineering by producing artifacts necessary for efficiency of application development. Therefore, domain engineering has to be tailored to CBSD process. The existing domain engineering methods dont elicit information necessary for Component-based Software Development process, such as, selecting and configuring appropriate components. Also, the existing domain analysis and design method do not represent objective analysis processes that extract and determine the properties of the domain, such as commonality and variability. In addition, the method that these domain information are explicitly reflected to the domain component and the domain architecture is deficient.

In this paper, the Component-Based Domain engineering method, which systematically defines, analyzes, and designs the domain as an effective reuse method to develop component-based software, is presented. The reusable components within the domain for mobile business, namely the common factors, are extracted objectively through the requirement analysis step for mobile business and should be continuously maintained, refined throughout the processes and reflected to the output of each step [7]. Through this process, the domain components with common factors can be identified. And with this base, domain architecture can be designed. Domain architecture for mobile business with the property of commonality and variability can be used for various systems that belong to the domain; thus, domain architecture has the flexibility that can reflect a variety of features of each system. Software reuse is possible through Mobile Business Domain Modeling (MBDM), which is presented in this study, and supports pre-cycle relationships of systematic replication by allowing implementation of reusable software.

2 Related Works

2.1 Software Product Line (SPL) for Mobile

Developing a SPL is very similar to developing a single system an intent to reuse large modules. However, SPLE puts a strong emphasis on documenting and modeling where the system can be varied [3]. This difference is very important in determining how the platform can be used and what can be changed to create the desired system. Figure 1 shows an overview of the process used to create a SPL. Domain Engineering is the process of defining how the platform will be able to change and generating artifacts that are common to the platform [8, 10, 11]. Application Engineering is the process of taking the artifacts that were produced in the domain engineering process and modifying the artifacts to create the new product. This is realized by adding or changing components from the platform.

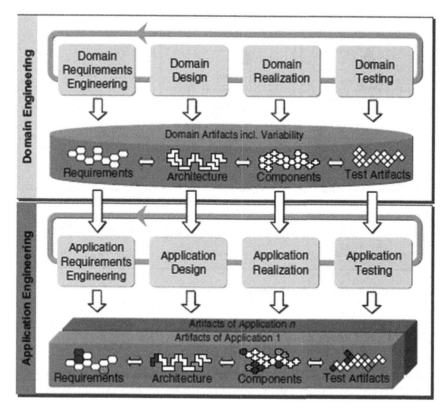

Fig. 1 Graphical representation of the process of developing a product line [9]

2.2 Variability and Documentation

Variability is the main component of a SPL and thusly needs to be documented well. There are a couple ways to capture variability in the product line; such as including variable features within standard UML diagrams [9]. The other method is using an Orthogonal Variability Model (OVM). "An orthogonal variability model is a model that defines the variability of a software product line. It relates the variability defined to other software development models such as feature models, use case models, design models, and test models" [1]. The OVM is a common place to store all information dealing with how and where a system can be changed or added to to create a new application. There are a few benefits to use the OVM over including variability into other artifacts, namely

- OVM are smaller and less complex than UML diagrams.
- Variability is defined consistently across all system artifacts [4].
- Communication is easier, due to a relatively easier to understand model compared to more complex UML models [9].

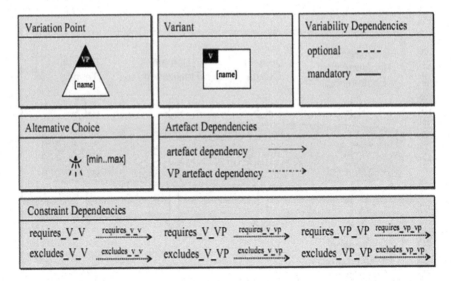

Fig. 2 Components found in an orthogonal variability model [1]

2.3 OVM, A Closer Look

Now that the benefits of using an OVM have been explained, we can explore how
the OVM is created. Figure 2 shows a list of common symbols that are found in the
OVM.

A Variation Point refers to a point that is able to change. It does not explicitly
state what the value can be, but that this is a point to change. A Variant is the specific
object that can be inserted at a variation point. Alternative choice is used to link
a Variation Point to a Variant. It also explains any constraints on the number of
variants that can be used at the Variation Point. Optional Variability Dependencies
show that variants are optional if the Variation Point is to be used in the Application.
A mandatory variability dependency requires that a variant be present if that Variation
Point is in use [1]. Constraint dependencies state dependencies between variants
and/or variation points. These are used to show if a variant requires another specific
variant or variation point to work. Artifact dependencies show a relationship between
the OVM and common UML diagrams. This link allows an easy way to show what
artifacts are dependent upon specific variants.

Figure 3 shows a simple example of using OVM to model a security system.
The diagram shows three different variation points and seven different variants. The
Security Variation Point has two distinct variants, a Basic package or Advanced
package. The Basic Package includes Keypad Door Locks and Motion Sensors,
which are variants of points Door Locks and Intrusion Detection respectively. The
inclusion of these features is captured with the requires constraint dependency.

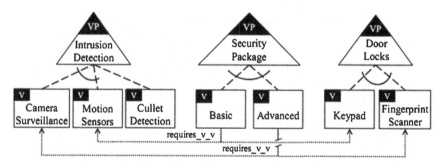

Fig. 3 OVM example of a security system [1]

3 Mobile Business Domain Modeling (MBDM)

The general mutual progression of Domain-driven Component-based Software development process presented in this study is as in Fig. 1.

Our process model for component-based software development explicitly considers reuse-specific activities, such as componential design, component identification, and component adaptation. It is comprised of seven major activities, starting with context comprehension and requirement analysis, continuing with the combination of componential design and component identification, component creation, component adaptation, and finally ending with component assembly. Throughout the process, explicitly stated domain artifacts—domain specifications, domain model, and domain architecture—are produced.

Component-based Domain Engineering depends on the component-based software development process.

In the first step of domain engineering, domain definition, the purpose of the domain is decided, and its scope is confirmed. In the domain modeling step, a domain model is obtained by analyzing the domain. Domain analysis has to identify the stable and the variable parts of the domain [5]. Based on this domain model, the domain components are identified, and the domain architecture is created. Our process model for domain engineering has an objective analysis activity in each step, i.e. generalization process. The generalization process is tasks that classify the properties of domain requirement, domain usecase, and domain component and transform these into reusable form according to the properties. The artifacts of each step are maintained and saved with interrelationships. They are reused as useful information during component-based software development.

In this paper, we suggest some domain engineering processes—domain definition, domain modeling, and domain design—for launching a study among the Domain-driven MBDM process that is represented in Fig. 4.

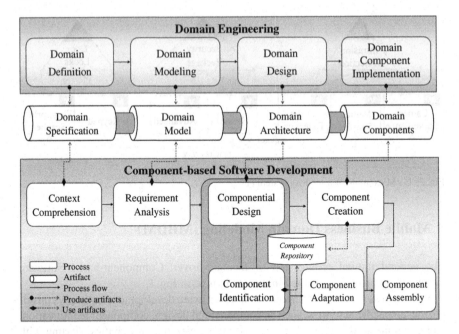

Fig. 4 Domain-driven MBDM Process

3.1 MBDM Domain Definition

The purpose of the Domain Definition step is to create domain specifications by bounding the domain scope and defining the domain purpose. In addition, requirements of domain are extracted from legacy and new systems in the domain and converted to generalized type reflecting properties—common, optional, variable.

3.1.1 Decide Domain Scope

As defined earlier, the domain is a collection of related systems, which can lead to vague interpretations, so it is imperative that ambiguities are made clear. If the scope of the domain is large, more systems can be contained in that domain, and it will be easy to contain new systems in the future. However, this leads to a reduction in commonality in the domain. Consequently, more commonality can be extracted in a domain of smaller scope that shows the detailed activities of domain engineering (Fig. 5).

Distinguish Domain External Stakeholder. Domain external stakeholder means people with interest in the functions provided by a domain. People who are interested in input or output of a domain or people who handle an external system related to the domain can be extracted as external stakeholders.

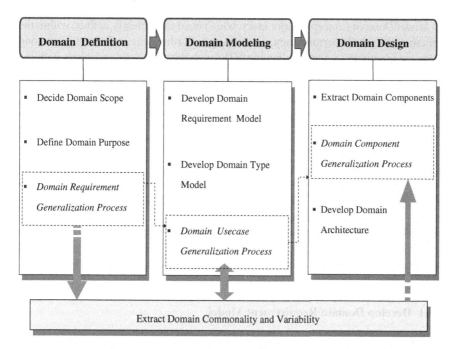

Fig. 5 Detailed activities of domain engineering

Define Domain Assumption. The domain assumption means pre-conditions that are to be satisfied by using components that are provided by the domain. Domain assumption performs a basic role to decide whether or not a system can be included in the domain in the initial step. Subsequently, it has influence on the decision of the domain's component properties that will be extracted later.

Describe Domain Environment. Domain environment is divided into domain external environment and domain internal environment. Domain external environment presents clearly the boundary of the domain by analyzing interaction between the domain and its external factors. Domain internal environment presents factors that should be distributed within the domain and its functions accordingly.

3.1.2 Define Domain Purpose

After the scope of the domain is set, a rough outline centering on the functionality of the domain is explained. Additionally, a domain concept schematic diagram is drawn outlining the domains business processes related to its purpose.

Describe Domain Purpose. The important function of the domain is described. It is an essential factor that all the systems belonging to the domain should have. Furthermore, it functions as a basis to decide whether the system should be included in the domain.

Model Domain Concept. Major tasks, which need to be clearly defined within the domain, and related terminology are extracted. Furthermore, relationships among these are identified in general drawings. Through these activities concepts within a domain are expressed.

3.2 Domain Modeling

The purpose of the Domain Modeling step is to analyze the domain and to develop the domain model composed of a domain requirement model and a domain type model with commonality and variability. A domain model captures the most important things—business objects or process and prepares variable things within the context of the domain. We use usecase analysis technique as an appropriate way to create such a model. The usecase leads to a natural mapping between the business processes and the requirements [2] by means of domain usecase description (Fig. 6).

3.2.1 Develop Domain Requirement Model

The domain requirement model expresses the requirements extracted from the domain by the usecase diagram of UML. This induces the analyzed primitive requirements to be bundle into a suitable unit.

Construct Domain Usecase Diagram. The actor is extracted from the domain stakeholder and the domain external environment. The requirements of such an actor and domain operations are extracted as a domain usecase.

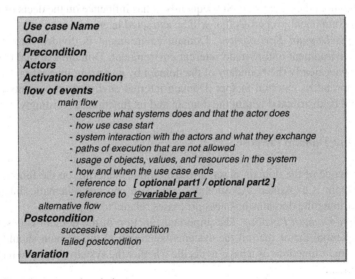

Fig. 6 Domain usecase description

Then a domain usecase diagram is drawn. The domain usecase is written with different levels of detail. The primitive requirements identified during the prior step should all be allocated to usecases. This provides an important link in terms of traceability between the artifacts. The domain usecase diagram should be modified to reflect the properties of domain usecase after the domain usecase generalization process (explained in Sect. 3.2.3).

Describe Domain Usecase Description. Domain usecase description is described in the items of Fig. 4 on each usecase. At this time, a reference appointment of variable requirement is presented with template marking—⊕.

3.2.2 Develop Domain Type Model

Based on the domain concept model produced during the domain definition step and the domain usecase description, the domain type model is developed by extracting detailed information and status that should be controlled by the system.

In this model, not only physical but also non-physical, such as a processor, can be a domain type. This allows common comprehension on the domain and enables the possibility of applying a consistent glossary to overall processes.

The domain type model is presented as a type of class diagram of UML. It defines the attributes of each domain type, and limitations of the model such as multiplicity of relationships.

3.2.3 Domain Usecase Generalization Process

A task to classify the properties of domain usecase and reconstruct the domain usecase according to these properties is defined as *domain usecase generalization process.* Properties of domain usecase are influenced by PRs properties.

Construct PR-Usecase Matrix. Create a PR-Usecase matrix to recognize the property of each usecase by referring to the domain usecase diagram and description and the PR-Context matrix. The usecase name, primitive requirement, and the property of primitive requirement are displayed in the matrix. Moreover, the primitive requirements that are contained in each usecase are analyzed. Figure 7 presents the PR-Usecase matrix.

Generalize Domain Usecase. When analyzing usecase, we can consider usecase conditions as the following; at this time, we can divide and rearrange usecases on their necessity. This is presented in Figs. 7 and 8 first in considering the usecase condition, a usecase contains primitive requirements, which does not overlapped with that of other usecases (① of Figs. 7 and 8). In this case, no re-arrangement is necessary. Second, the primitive requirement is spread over to many usecases (② of Figs. 7 and 8). In this case, separate commonly overlapped primitive requirements, make it an independent usecase, and connect it to include-relationships. Third, a usecase includes variable primitive requirements (③ of Figs. 7 and 8). In this case,

Usecase \ Req.	property	Usecase1	Usecase2	Usecase3	Usecase4	...	Usecase k
PR1	C	O					
PR 2	C	O					
PR k	:	①	O				
PR m			O	O			④
PR n			②	O			O
:					O		
PR k	V				O		
PR m	P				③	O	O

C: Common PR   V: Variable PR   P: Optional PR

Fig. 7 PR-usecase matrix

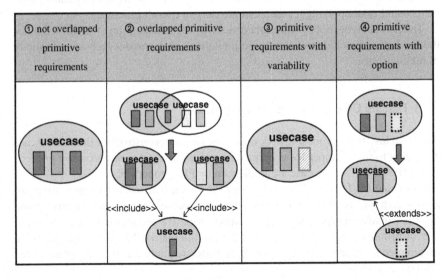

① not overlapped primitive requirements	② overlapped primitive requirements	③ primitive requirements with variability	④ primitive requirements with option

Fig. 8 Property identification process from PR-usecase matrix

a confirmation on the possibility of whether variable primitive requirements can be separated and created as independent usecase is addressed.

If possible, they are separated. If not, they are maintained as involved in a usecase and a variable point is stored. Finally, a usecase includes optional primitive requirements (④ of Figs. 7 and 8). In this case, the optional primitive requirements are separated and connected to the extend-relationship shown in (Fig. 9) which is the domain component extraction standard for mobile.

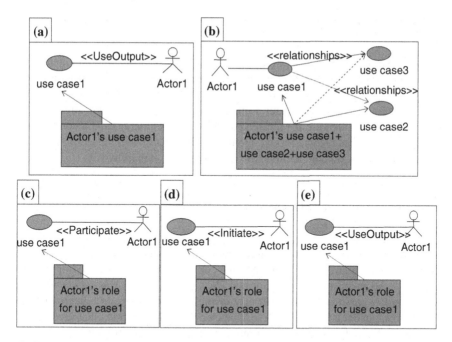

Fig. 9 Domain component extraction standard for mobile

The usecases that were reorganized by this process are classified by the properties as follows:

- **Common Usecase**—When the usecase has primitive requirements, which must exist within the domain, it is classified as a common usecase and represents an important process in the system.
- **Variable Usecase**—When the usecase is composed of variable primitive requirements, this is classified as a variable usecase. It means a usecase with requirements that exist in each specific application of a domain but can be variable. Mainly, it tends to appear overlapping in many usecases. In case it was divided into an independent usecase through the PR-Usecase matrix analysis, it belongs to this class.
- **Optional Usecase**—It represents the usecase that doesnt always need to exist when handling a process in the system; this corresponds to a usecase composed of optional primitive requirements.
- **Usecase with variables**—Usecase with variables—When primitive requirements with variation are difficult to be separated independently, this is involved in the usecase. Even though this cannot be divided separately, it can be used when identifying a domain component of the next step and draw the component interaction diagram at the domain design step by classifying this status.

3.3 Domain Design

The purpose of the Domain Design step is to identify the domain components and to develop the domain architecture. The domain component, which is different from the physical component that can be deployed immediately during software development, is defined as a service central unit package of platform independent logical level. Domain architecture is represented out of the identified domain components in a concrete and analyzable format. Domain architecture is different from software architecture in that domain architecture must allow for variability [7].

3.3.1 Domain Component Generalization Process

Each extracted domain component executes a relationship based on usecase, and is divided by its properties of commonality, optional, and variability in review of the PR-Usecase matrix. In this process, the components are reorganized upon their necessity, and an abstraction is performed. We call this process *"domain component generalization process."* If the property of the usecase is pure, it reflects that the domain component is also pure. Pure character means the property that usecase has is composed of only one among commonality, variability, or optional.

- **Common Domain Component**—Common Domain Component is extracted from common usecase.
- **Optional Domain Component**—Optional Domain Component is extracted from optional usecase.
- **Variable Domain Component**—Variable Domain Component is extracted from variable usecase.
- **Domain Component with Hole**—If the usecase has mixed property (not pure), namely a usecase with variable, a domain component is converted to a more general type. The domain component is abstracted by replacing the variable part with $\ll$ Hole $\gg$, $\ll$ Hole $\gg$ becomes an area that fills contents differently by each production of the application component.

3.3.2 Develop Domain Architecture

Domain architecture presents the structure of domain components, interaction between domain components in multiple views, and specifications of the domain component. Domain architecture should be independent of any specific technology or set of developmental tools. Domain architecture should reflect properties such as commonality, variability and optional that were initialized from the requirement analysis step, refined, and maintained. Such features allow part of the architecture to be divided and replaced according to the components property when creating the component-based software development. So, malleable architecture can be created.

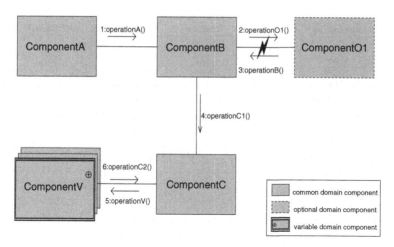

Fig. 10 Domain component interaction view

Domain Component Interaction View. Domain Component Interaction View represents interactions between domain components that perform specific requirements. In addition, a domain component interface is extracted by analyzing operations between components. Domain Component Interaction View is presented by using an Interaction diagram, and component interface is described by using class notation. Figure 10 presents Domain Component Interaction View. One interface is a set of externally visible operations provided by a component. The property of the operation is defined based on the property of the domain component that was extracted. That is, all operations that variable and optional domain component provide are defined as variable and optional operations respectively. But, variable operation may be additionally identified from common domain component by refining the function of the domain component. Such operation is presented as ≪ v.p ≫ (variant point) when describing interface. The interface of common domain component has not only provided operations but also required operations information. At this time, the properties of a required operation can be identified by analyzing the interaction of the domain components that have a different property. For example, a required operation of the ComponentB, operationO1 is defined as optional property based on ComponentO1 with optional property as shown in Fig. 10. Details on interface will be described in domain component specifications, and expected cases on variable operations will also be described together. In this view, variable domain components can be templated and optional components can be pruned in this view. Through these processes, the domain component interaction view becomes the analysis model that has common elements only, which is defined as Commonality Analysis Model in this paper. The Commonality Analysis Model presents an execution view of most basic steps of component based software development.

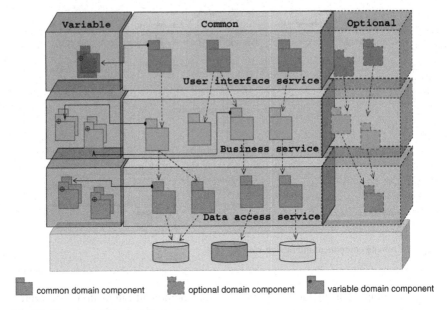

common domain component   optional domain component   variable domain component

Fig. 11 Domain development view

Domain Development View. All computer applications have three general areas of functionality; User services, Business services and Data services. Domain structure can be divided into common, variable and optional parts by the features of the domain.

In the domain development view, the domain structure is divided and presented in namely 2nd dimension layers through logical partitioning of functionality as horizontal and property division of the domain as vertical. Not only does the systemized view allow for independent performance and quick change at each step, but also is becomes the foundation for various physical partitioning (deployment alternatives) such as 2 tier or 3 tier, n tier, and Web-enabled applications. The vertically divided view determines the optional component easily by the applications specific factor, and easily supports modifying the variable component, so it covers various systems that belong to the domain. Figure 11 represents the Domain Development View of a 2nd dimension division.

Domain Component Specifications. Domain Component Specifications describes the purpose and interfaces of a component and furnishes information about what the component does and how it can be used. A deployable component, which is developed using a Domain Component Specifications, can differ in granularity according to applications. Hence, we will describe the related functions as interface and supplement required interfaces to effectively support variable granularity of the component. In this way when interfaces are developed independently, required interfaces can be recognized easily. Also the Domain Component property is explicitly represented using a 'type' tag in the interface. The 'type' tag can have common, variable, or optional values. If the 'type' tag has a variable or optional value, it can be described

UseCase Primitive requirement	Property	Login	Manage a News	Manage a member	Manage a scrapbook
Login	C	o			
Logout	C	o		Optional Usecase	
Modify the member information	C	o			
Unregister	C	o			
Search the news	C		o		
Show the news	C		o		
Show the previous news in a new way	Vc		o		
Write the news	C		o		
Delete the news	C		o		
Changed service	Vp		o		
Send e-mail about the news to the member	Vc		o		
:					
Modify the member information	C			o	
:				o	
Delete the news in scrapbook	P				o
Read the news in scrapbook	P				o
:					

C : common PR   P : optional PR   Vc : variable in common PR   Vp : variable in optional PR

Fig. 12 PR-usecase matrix of news information storage domain

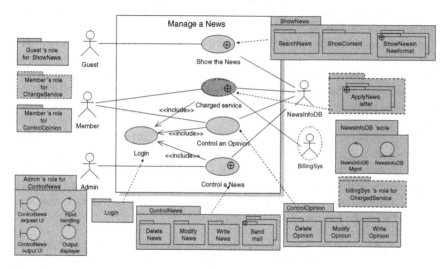

Fig. 13 Domain components extracted from manage a news domain usecase

as a predictable case by a 'rule' tag in the interface. Figure 10 presents Domain Component Specifications PR-usecase matrix of news information storage domain shown in Fig. 12 and Fig. 13 shows the Domain components extracted from manage a news domain usecase.

4 Conclusion and Future Work

In this study, which is different from the existing methods, processes were suggested for domain analysis and design method that have mutual action suited to component-based software development.

Namely, in the existing study, it could not obtain information on procedures to assemble components in consideration of their relationship using architecture and recognition of domain components. Therefore, this study recognized and selected components that were required during the development of component base software. Furthermore, a connected relationship between the components and the interface information through domain engineering processes were calculated which supported the component base software development process.

Also, this study deducted a method to find commonality and variability, which is necessary to extract information with objectivity through the generalization processes while existing studies depended solely on experience and intuition by a domain specialist. In addition, this information was relocated into matrix form to be maintained, refined, and used in each step to find common usecase and common domain component.

This study reflected such features into the shape of the domain architecture, created a malleable architecture that can be partially separated architecture, and replaced them by the property of the components during component based software development.

Future studies will progress in two directions. We will review the possibility to implement a domain component by using domain architecture and study ideas and technologies that can be applied. Also we will study a process that can be implemented for the development of component based software by using a proposed domain analysis and design method. Hereby, the general study process that was proposed from the beginning will be completed.

Acknowledgments This work (Grants No. C0124408) was supported by Busi-ness for Cooperative R&D between Industry, Academy, and Research Institute funded Korea Small and Medium Business Administration in 2013.

References

1. Bockle, G., Pohl, K., van der Linden, F.: Software Product Line Engineering. Springer, Germany (2010)
2. Creps D., Klingler, C., Levine, L., and Allemang, D.: Organization Domain Modeling (ODM) Guidebook Version 2.0, Software Technology for Adaptable, Reliable Systems (STARS) (1996).
3. A framework for software product line practice, version 5.0. Retrieved from http://www.sei.cmu.edu/productlines/frame_report/testing.htm (2009)
4. Jaring, M., Krikhaar, R.L., Bosch, J.: Modeling variability and testability interaction in software product line engineering mode. In: Proceedings of the Composition-

Based Software Systems, http://ieeexplore.ieee.org.ezproxy.uwplatt.edu/stamp/stamp.jsp? tp=&arnumber=4464016 2008, doi:10.1109/ICCBSS.2008.9

5. Kang, K.C.: Feature-oriented domain analysis for software reuse, Ioint Conference On Software Engineering (1993) pp. 389–395.
6. Kang, K.C., Kim, S., Lee J., and Kim, K.: FORM: a feature-oriented reuse method with domain specific reference architectures, Pohang University of Science and Technolog (POSTECH) (1998).
7. Klingler, C.D.: DAGAR: a process for domain architecture definition and asset implementation. In: Proceedings of ACM TriAda (1996).
8. Li, D., Chang, C.K.: Initiating and institutionalizing software product line engineering: from bottom-up approach to top-down practice. In: Proceedings of the Computer Software and Applications Conference, http://ieeexplore.ieee.org.ezproxy.uwplatt.edu/stamp/stamp. jsp?tp=&arnumber=5254280 2009, doi:10.1109/COMPSAC.2009.17
9. Metzger, A., Pohl, K.: Variability management in software product line engineering. In: Proceedings of the 29th International Conference on Software Engineering, http://ieeexplore. ieee.org.ezproxy.uwplatt.edu/stamp/stamp.jsp?tp=&arnumbe=4222738 2007, doi:10.1109/ ICSECOMPANION.2007.83
10. Schaefer, I., Hahnle, R.: Formal methods in software product line engineering. Computer, 44(2), http://ieeexplore.ieee.org.ezproxy.uwplatt.edu/stamp/stamp.jsp?tp=&arnumber =5713307(2011)
11. SEI in Carnegie Mellon University: Domain Engineering and Domain Analysis, URL:http:// www.sei.cmu.edu/str/descriptions/dade.html

A New Modified Elman Neural Network with Stable Learning Algorithms for Identification of Nonlinear Systems

Fatemeh Nejadmorad Moghanloo, Alireza Yazdizadeh
and Amir Pouresmael Janbaz Fomani

Abstract In this paper a new dynamic neural network structure based on the Elman Neural Network (ENN), for identification of nonlinear systems is introduced. The proposed structure has feedbacks from the outputs to the inputs and at the same time there are some connections from the hidden layer to the output layer, so that it is called as Output to Input Feedback, Hidden to Output Elman Neural Network (OIFHO ENN). The capability of the proposed structure for representing nonlinear systems is shown analytically. Stability of the learning algorithms is analyzed and shown. Encouraging simulation results reveal that the idea of using the proposed structure for identification of nonlinear systems is feasible and very appealing.

Keywords Elman Neural Network · OIFHO ENN · Nonlinear System Identification

1 Introduction

In broad terms, the ultimate goal of system identification is to obtain a mathematical model whose output matches the output of a dynamic system for a given input. The solution to the exact matching problem, in general, is extremely difficult. Consequently, for practical reasons the original problem is relaxed to development of a model whose output can be made "as close as possible" to the output of the considered dynamic system. Different methods have been developed in recent years for linear/nonlinear system identification. A common characteristic of most of these methods is the use of a parameterized model where parameters are adjusted based on the minimization of a norm of the output identification error. These methods can be

F. Nejadmorad Moghanloo (✉) · A. Yazdizadeh · A. Pouresmael Janbaz Fomani
Department of Electrical Engineering, Abbaspour College of Technology,
Shahid Beheshti University, 16765-1719, Tehran, Iran
e-mail: f.nejadmorad@gmail.com

A. Yazdizadeh
e-mail: Alireza@pwut.ac.ir

A.P.J. Fomani
e-mail: a.pouresmael@gmail.com

© Springer International Publishing Switzerland 2015   171
R. Lee (ed.), *Computer and Information Science*, Studies in Computational
Intelligence 566, DOI 10.1007/978-3-319-10509-3_13

classified into two main categories, namely, conventional and neural network-based methods [1]. Conventional methods are based on well established linear system theory and recently developed nonlinear system techniques. In most of existing methods, under certain conditions, desired characteristics such as convergence of the output error (identification error) to zero and stability of the identifier are shown analytically. The main disadvantage of these methods is that they are generally applicable and extendible to only a special class of nonlinear systems. In order to generalize these results to arbitrary classes of nonlinear systems, restrictive knowledge about the system is required [1].

Fortunately, the characteristics of the Artificial Neural Network (ANN) approach, namely nonlinear transformation, provide effective techniques for system identification, especially for non-linear systems. The ANN approach has a high potential for identification applications because: (1) it can approximate the nonlinear input–output mapping of a dynamic system; (2) it enables to model the complex system's behavior through training, without a priori information about the structures or parameters of systems. Due to these characteristics, there has been a growing interest, in recent years, in the application of neural networks to dynamic system identification and control [2–5].

1.1 Literature Review

Elman neural network (ENN) is a partial recurrent network model proposed by Elman in 1990 [6]. It lies somewhere between a classic feedforward perception and a pure recurrent network. The feedforward connection consists of the input layer, hidden layer, and output layer, in which the weights connecting two neighboring layers are variables. In contrast to the classical feedforward neural networks, the back forward connection employs context layer that is sensitive to the history of input data, therefore, the connections between the context layer and the hidden layer are fixed. Furthermore, since dynamic characteristics of Elman network is provided by internal connections, it does not need to use the state as input or training signals, which makes ENN superior to static feedforward network and is widely used in dynamic system identification [7].

There has been much research interest in Elman Neural Network [2, 8–15]. Elman Neural Network has been applied to dynamic system identification and financial prediction in [8, 10], respectively. A modified Elman Neural Network has been proposed by [4] because it was found that the basic Elman network trained by the standard Backpropagation (BP) algorithm was able to model only first-order dynamic systems. The performance of Elman's RNN has shown by means of two different applications in [14]. Song [15] focuses on the real-time online learning of an extended training algorithm for Elman Neural Network with a new Multiple-Input–Multiple-Output (MIMO) adaptive dead zone scheme and guaranteed weight convergence. Hsu [16] proposes an Elman-based self-organizing RBF Neural Network (ESRNN) for online approximation of the unknown nonlinear system dynamics based on a Lyapunov function and an Adaptive Backstepping Elman-based Neural Control (ABENC) system to eliminate the effect of the approximation error. Pham has described the dynamic BP (DBP) algorithm in [8] which is proper for training the basic Elman Neural Network and shows that the modified Elman Neural Network is an approximation of the Elman Neural Network trained by DBP. Pham has clarified why the modified Elman Neural Network can model higher-order dynamic systems. In [11] OHF and OIF Elman Neural Networks are presented for identification and control of ultrasonic motor. A hybrid Elman- NARX Neural Network is presented by [12] to analyze and predict chaotic time series. A new recurrent Neural Network based on the original Elman Neural Network is introduced in [2] to improve the resolution ratio of Elman Neural Network. Yuan Cheng has presented a new modified Elman Neural Network to improve the dynamic characteristics of the original Elman Neural Network [17]. A novel EMD–ENN approach, a hybrid of Empirical Mode Decomposition (EMD) and Elman neural network (ENN), is presented in [5] to forecast the wind speed. In this study, first, the original wind speed dataset are decomposed into sub-series with EMD and then each sub-series are forecasted using an Elman Neural Network model. The forecasted values of original wind speed are calculated by the sum of the predicted values of every sub-series.

1.2 Contributions

The contributions of this paper are as follows:

- To present a new modified Elman neural network in which covers four classes of nonlinear systems. The feedback information from all layers can improve dynamic characteristics and convergence speed of the new modified Elman neural network. It possesses comparatively higher learning capability and convergence speed.
- To analysis the stability of the learning rates.
- To compare numerical results obtained through the proposed approach of this paper with ones achieved from other modified Elman neural networks reported in the literature.

1.3 Paper Organization

The organization of this paper is as follows. Section 2, introduces the proposed modified Elman neural network, namely OIFHO ENN for identification of general classes of nonlinear systems and develops the dynamic recurrent back-propagation algorithm for the purposed new modified Elman neural network. In Sect. 3, to guarantee the fast convergence, the optimal adaptive learning rates are also derived in the sense of discrete-type Lyapunov stability. Simulation results are presented in Sect. 4. Section 5 uses different norms of error, namely, Mean Square Error (MSE), Root Mean Square Error (RMSE) and Normalized Mean Square Error (NMSE) to evaluate the performance of the new modified Elman Neural Network structure proposed in Sect. 2 in comparison with the OHF and the OIF structures.

2 New Modified Elman Neural Network (OIFHO ENN)

In the original Elman neural network, the hidden layer neurons are fed by the outputs of the context neurons and the input neurons. Context neurons are known as memory units as they store the previous outputs of hidden neurons. Since a typical Elman neural network only employs the hidden context nodes to diverse message, it has low learning speed and convergence precision. On the other hand, in the proposed modified Elman neural network, the feedback of the output layer is taken into account; therefore better learning efficiency can be obtained. Moreover, to make the neurons sensitive to the history of input data, self connections of the context nodes and output feedback node are added. Thus, the proposed modified Elman neural network combines the ability of dealing with nonlinear problems and can effectively improve the convergence precision and reduce learning time.

Figure 1 depicts our proposed new modified Elman neural network, namely, Output to Input Feedback, Hidden to Output Elman Neural Network (OIFHO ENN) that is presented based on the Elman neural network. The OIFHO ENN possesses self-feedback links with fixed coefficient α, β and γ in the context nodes. The feedback information from all layers can improve dynamic characteristics and convergence speed of the new modified Elman neural network. In order to compare the speed of convergence of the proposed method with other method, we try several benchmark examples in Sect. 5.

The Input–output equation of OIFHO ENN is:

$$
\begin{aligned}
Y(k) &= g\left(W^4(k)X_c(k) + W^5(k)X(k)\right) \\
&= W^4(k)(\gamma X_c(k-1) + X(k-1)) + W^5(k)X(k) \\
&= N_f[U(k), U(k-1), \ldots, Y(k-1), Y(k-2), \ldots]
\end{aligned}
\tag{1}
$$

$$
y_{c,l}(k) = \beta y_{c,l}(k-1) + y_l(k-1), \quad l = 1, \ldots, n
\tag{2}
$$

$$
x_{c,k}(k) = \gamma x_{c,k}(k-1) + x_k(k-1), \quad k = 1, \ldots, n
\tag{3}
$$

where $y_{c,l}$ is the output of the lth output context unit, $x_{c,k}$ is the output of the kth hidden context unit and β $(0 < \beta \leq 1)$ and γ $(0 < \gamma \leq 1)$ are the self-feedback coefficients. It is a type of recurrent neural networks with different layers of neurons, namely: input nodes, hidden nodes, output nodes and context nodes. The input and output nodes interact with the outside environment, whereas the hidden and context nodes do not. The context nodes are used only to memorize previous activations of the hidden nodes and the output nodes. The feed–forward connections are modifiable, whereas the recurrent connections are fixed.

If we assume that there are r nodes in the input layer, n nodes in the hidden layer and the hidden context layer and m nodes in the output layer and the output context layer, then the input u is an r dimensional vector and the output x of the hidden layer and the output $x_{c,k}$ of the hidden context nodes are n dimensional vectors, where the output y of the output layer and the output $y_{c,l}$ of the output context nodes are m dimensional vectors, and the weights W^1, W^2, W^3, W^4, and W^5 are the weights between hidden layer and input layer, input layer and output context layer, hidden layer and output context layer, output layer and hidden context layer and output layer and hidden layer and are $n \times r$, $r \times m$, $n \times m$, $m \times n$, and $m \times n$ dimensional matrices, respectively.

The mathematical model of the new modified Elman Neural Network can be described as follows:

$$y\,(k) = g\left(W^4 x_c\,(k) + W^5 x\,(k)\right) \tag{4}$$

$$x_c\,(k) = \gamma x_c\,(k-1) + x\,(k-1) \tag{5}$$

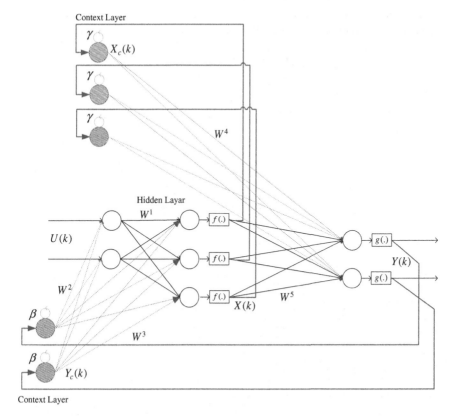

Fig. 1 Architecture of the new modified Elman neural network (OIFHO ENN)

$$x(k) = f\left(W^1\left(W^2 y_c(k) + u(k)\right) + W^3 y_c(k)\right) \tag{6}$$

$$y_c(k) = \beta y_c(k-1) + y(k-1) \tag{7}$$

$f(x)$ is often taken as the sigmoid function:

$$f(x) = \frac{1}{1 + e^{-x}} \tag{8}$$

and $g(x)$ is often taken as a linear function, that is:

$$y(k) = W^4 x_c(k) + W^5 x(k) \tag{9}$$

We may define a norm of error as:

$$E(k) = \frac{1}{2}(y_d(k) - y(k))^T (y_d(k) - y(k)) \tag{10}$$

By differentiating E with respect to W^1, W^2, W^3, W^4, and W^5, according to the gradient descent method, we obtain the following equations:

$$W(k+1) = \Delta W(k) + W(k)$$

$$\Delta w_{ij}^5(k) = -\eta_5 \frac{\partial(e_i(k))}{\partial w_{ij}^5(k)} = \eta_5 . \delta_i^o(k)(x_j(k) + w_{ij}^5(k)\frac{\partial x_j(k)}{\partial w_{ij}^5(k)})$$
$$(i = 1, \ldots, m), (j = 1, \ldots, n) \tag{11}$$

$$\frac{\partial x_j(k)}{\partial w_{ij}^5(k)} = f_j'(.)\left(\sum_{q=1}^{r} w_{jq}^1(k).A_5 + \sum_{l=1}^{m} w_{jl}^3(k).B_5\right) \tag{12}$$

where

$$A_5 = \sum_{l=1}^{m} w_{ql}^2(k)(\gamma \frac{\partial y_{c,l}(k-1)}{\partial w_{ij}^5(k)} + (x_j(k-1) + w_{ij}^5(k)\frac{\partial x_j(k-1)}{\partial w_{ij}^5(k)}))$$

$$B_5 = \left(\gamma \frac{\partial y_{c,l}(k-1)}{\partial w_{ij}^5(k)} + (x_j(k-1) + w_{ij}^5(k)\frac{\partial x_j(k-1)}{\partial w_{ij}^5(k)})\right)$$

$$(i = 1, \ldots, m), (j = 1, \ldots, n)$$

$$\Delta w_{ih}^4(k) = -\eta_4 \frac{\partial(e_i(k))}{\partial w_{ih}^4(k)} = \eta_4 \delta_i^o(k) \left(x_h(k-1) + \gamma \frac{\partial y_i(k-1)}{\partial w_{ih}^4(k)} \right) \tag{13}$$
$$(h = 1, \ldots, n), (i = 1, \ldots, m)$$

$$\Delta w_{jl}^3 = \eta_3 \frac{\partial(x_j(k))}{\partial w_{jl}^3} . \delta_j^h(k) = \eta_3 \sum_{j=1}^n w_{ij}^5 \delta_i^o(k) \frac{\partial x_j(k)}{\partial w_{jl}^3}$$
$$= \eta_3 \sum_{j=1}^n w_{ij}^5 \delta_i^o(k) \left(f_j'(.) y_l(k-1) + \beta \frac{\partial x_j(k-1)}{\partial w_{jl}^3} \right) \tag{14}$$
$$(j = 1, \ldots, n), (q = 1, \ldots, r), (p = 1, \ldots, m)$$

$$\Delta w_{ql}^2(k) = \eta_2 \frac{\partial(x_j(k))}{\partial w_{ql}^2(k)} . \delta_j^h(k)$$
$$= \eta_2 \sum_{q=1}^r \sum_{j=1}^n w_{jq}^1(k) w_{ij}^5(k) f_j'(.)$$
$$\times \left(\begin{matrix} \gamma \frac{\partial y_{c,l}(k-1)}{\partial w_{ql}^2(k)} \\ + (\sum_{h=1}^n w_{ih}^4(k) \frac{\partial x_{c,h}(k-1)}{\partial w_{ql}^2(k)} + \sum_{j=1}^n w_{ij}^5(k) \frac{\partial x_j(k-1)}{\partial w_{ql}^2(k)}) \end{matrix} \right) . \delta_i^o(k)$$
$$(l = 1, \ldots, m) \tag{15}$$

$$\Delta w_{jq}^1(k) = \eta_1 \frac{\partial(x_j(k))}{\partial w_{jq}^1(k)} . \delta_j^h(k) = \eta_1 \sum_{j=1}^n w_{ij}^5(k) f_j'(.) \frac{\partial(x_j(k))}{\partial w_{jq}^1(k)} . \delta_i^o(k)$$
$$\frac{\partial x_j(k)}{\partial w_{jq}^1(k)} = (u_q(k) + \sum_{l=1}^m w_{ql}^2(k) y_{c,l}(k)) + (\sum_{q=1}^r w_{jq}^1(k).A_1 + \sum_{l=1}^m w_{jl}^3(k).B_1)$$
$$\tag{16}$$
$$(q = 1, \ldots, r)$$

where

$$A_1 = \sum_{l=1}^m w_{ql}^2(k)(\gamma \frac{\partial y_{c,l}(k-1)}{\partial w_{jq}^1(k)} + (\sum_{h=1}^n w_{ih}^4(k) \frac{\partial x_{c,h}(k-1)}{\partial w_{jq}^1(k)} + \sum_{j=1}^n w_{ij}^5(k) \frac{\partial x_j(k-1)}{\partial w_{jq}^1(k)})))$$
$$B_1 = (\gamma \frac{\partial y_{c,l}(k-1)}{\partial w_{jq}^1(k)} + (\sum_{h=1}^n w_{ih}^4(k) \frac{\partial x_{c,h}(k-1)}{\partial w_{jq}^1(k)} + \sum_{j=1}^n w_{ij}^5(k) \frac{\partial x_j(k-1)}{\partial w_{jq}^1(k)}))$$

which form the learning algorithm for the OIFHO ENN, where η_1, η_2, η_3, η_4, and η_5 are learning rates of W^1, W^2, W^3, W^4, and W^5, respectively, and

$$\delta_i^o(k) = (y_{d,i}(k) - y_i(k)) g_i'(.) \tag{17}$$

$$\delta_j^h(k) = \sum_{i=1}^{m} w_{ji}^5(k)\delta_i^o(k)f_j'(.) \tag{18}$$

$$\delta_q^i(k) = \sum_{j=1}^{n}\sum_{i=1}^{m} w_{jq}^1(k)w_{ji}^5(k)\delta_i^o(k)f_j'(.) \tag{19}$$

if $g(x)$is taken as a linear function, then $g'(.) = 1$.

3 Convergence of Output to Input Feedback, Hidden to Output Elman Neural Network (OIFHO ENN)

The update rules in Eqs. (12–16) need appropriate choice of the learning rates. For the learning rate with a small value, the convergence can be guaranteed, but the speed of convergence is very slow. On the other hand, if the value of the learning rate is too large, the algorithm will become unstable [2]. In order to train neural networks efficiently, we propose five criterions of selecting proper learning rates for the dynamic back propagation algorithm adaptively based on the discrete-type Lyapunov stability analysis. The following theorems give sufficient conditions for the convergence of OIFHO ENN. Suppose that the modification of the weights of by Eqs. (12–16). For the convergence of OIFHO ENN we have the following theorems.

Theorem 1 *The stable convergence of the update rule on* W^1 *is guaranteed if the learning rate* $\eta_1(k)$ *satisfies the following condition:*

$$0 < \eta_1(k) < \frac{8}{nr\left|\max_{ij}(w_{ij}^5(k))\right|\left\|\left(\max_q\left|u_q(k)\right| + \max_q\left|\sum_{p=1}^{m} w_{qp}^2 y_{c,p}(k)\right|\right)\right\|} \tag{20}$$

Proof Define the Lyapunov energy function as follows:

$$E(k) = \frac{1}{2}\sum_{i=1}^{m} e_i^2(k) \tag{21}$$

where

$$e_i(k) = y_{d,i}(k) - y_i(k) \tag{22}$$

and consequently, we can obtain the modification of the Lyapunov energy function

$$\Delta E(k) = E(k+1) - E(k) = \frac{1}{2}\sum_{i=1}^{m}[e_i^2(k+1) - e_i^2(k)] \tag{23}$$

the error during the learning process can be represented as

$$e_i(k+1) = e_i(k) + \sum_{j=1}^{n}\sum_{q=1}^{m}\frac{\partial e_i(k)}{\partial w_{jq}^1}\Delta w_{jq}^1 = e_i(k) - \sum_{j=1}^{n}\sum_{q=1}^{m}\frac{\partial y_i(k)}{\partial w_{jq}^1}\Delta w_{jq}^1 \quad (24)$$

therefore

$$\Delta E(k) = \frac{1}{2}\sum_{i=1}^{m}e_i^2(k)\left[\left(1 - \eta_1(k)[\frac{\partial y_i(k)}{\partial W^1}]^T[\frac{\partial y_i(k)}{\partial W^1}]^2\right)^2 - 1\right]$$

$$= \frac{1}{2}\sum_{i=1}^{m}e_i^2(k)\left[\left(1 - \eta_1(k)\left\|\frac{\partial y_i(k)}{\partial W^1}\right\|^2\right)^2 - 1\right] = -\sum_{i=1}^{m}e_i^2(k)\beta_i^1(k)$$

$$(25)$$

where

$$\beta_i^1(k) = \frac{1}{2}\left[1 - \left(1 - \eta_1(k)\left\|\frac{\partial y_i(k)}{\partial W^1}\right\|^2\right)^2\right]$$

$$= \frac{1}{2}\eta_1(k)\left\|\frac{\partial y_i(k)}{\partial W^1}\right\|^2\left(2 - \eta_1(k)\left\|\frac{\partial y_i(k)}{\partial W^1}\right\|^2\right) \quad (26)$$

We have

$$\left|\frac{\partial y_i(k)}{\partial w_{jq}^1}\right| = \left|\frac{\partial y_i(k)}{\partial x_j(k)}\cdot\frac{\partial x_j(k)}{\partial w_{jq}^1}\right| = \left|w_{ij}^5(k).f_j'(.).(u_q(k) + \sum_{p=1}^{m}w_{qp}^2 y_{c,p}(k))\right| \quad (27)$$

$$(i = 1,\ldots,m : \; j = 1,\ldots,n : \; q = 1,\ldots,r)$$

then

$$\left|\frac{\partial y_i(k)}{\partial w_{jq}^1}\right| \le \frac{1}{4}\left|\max_{ij}(w_{ij}^5(k))\right|\left\|\left(\max_q|u_q(k)| + \max\left|\sum_{p=1}^{m}w_{qp}^2 y_{c,p}(k)\right|_q\right)\right\| \quad (28)$$

$$(i = 1,\ldots,m : \; j = 1,\ldots,n : \; k = 1,\ldots,n : \; q = 1,\ldots,r)$$

then

$$\left\|\frac{\partial y_i(k)}{\partial W^1}\right\| \le \sqrt{\frac{nr}{4}\left|\max_{ij}(w_{ij}^5(k))\right|\left\|\left(\max_q|u_q(k)| + \max\left|\sum_{p=1}^{m}w_{qp}^2 y_{c,p}(k)\right|_q\right)\right\|}$$

$$(i = 1,\ldots,m : \; j = 1,\ldots,n : \; k = 1,\ldots,n : \; q = 1,\ldots,r)$$

$$(29)$$

and we have

$$0 < \eta_1(k) < \frac{8}{nr \left| \max_{ij}(w_{ij}^5(k)) \right| \left\| \left(\max_q |u_q(k)| + \max \left| \sum_{p=1}^m w_{qp}^2 y_{c,p}(k) \right|_q \right) \right\|} \tag{30}$$

We have $\beta_i^1(k) > 0$, then from Eq.(25) we obtain $\Delta E(k) < 0$. According to the Lyapunov stability theory, this shows that the training error will converge to zero as $t \to \infty$. This completes the proof.

Theorem 2 *The stable convergence of the update rule (15) on W^2 is guaranteed if the learning rate $\eta_2(k)$ satisfies the following condition:*

$$0 < \eta_2(k) < \frac{8}{mn \left| \max_{ij}(w_{ij}^5(k)) \right| \left\| \max_{jq}(w_{jq}^3(k)) \right\| \left| \max_p y_{cp}(k) \right|} \tag{31}$$

Proof The error during the learning process can be expressed as

$$e_i(k+1) = e_i(k) + \sum_{j=1}^n \sum_{q=1}^r \frac{\partial e_i(k)}{\partial w_{jq}^2} \Delta w_{jq}^2 = e_i(k) - \sum_{j=1}^n \sum_{q=1}^r \frac{\partial y_i(k)}{\partial w_{jq}^2} \Delta w_{jq}^2 \tag{32}$$

therefore

$$\Delta E(k) = \frac{1}{2} \sum_{i=1}^m e_i^2(k) \left[\left(1 - \eta_2(k)[\tfrac{\partial y_i(k)}{\partial W^2}]^T[\tfrac{\partial y_i(k)}{\partial W^2}] \right)^2 - 1 \right]$$
$$= \frac{1}{2} \sum_{i=1}^m e_i^2(k) \left[\left(1 - \eta_2(k)\|\tfrac{\partial y_i(k)}{\partial W^2}\|^2 \right)^2 - 1 \right] = - \sum_{i=1}^m e_i^2(k)\beta_i^2(k) \tag{33}$$

where

$$\beta_i^2(k) = \frac{1}{2} \left[1 - \left(1 - \eta_2(k) \left\| \frac{\partial y_i(k)}{\partial W^2} \right\|^2 \right)^2 \right]$$
$$= \frac{1}{2}\eta_2(k) \left\| \frac{\partial y_i(k)}{\partial W^2} \right\|^2 \left(2 - \eta_2(k) \left\| \frac{\partial y_i(k)}{\partial W^2} \right\|^2 \right) \tag{34}$$

Notice that the activation function of the hidden neurons in the modified Elman neural network is the sigmoidal type, we have

$$\left|\frac{\partial y_i(k)}{\partial w_{jq}^2}\right| = \left|w_{ij}^5(k)w_{jq}^3(k)f_j'(.)y_{cp}(k)\right|$$

$$\leq \frac{1}{4}\left|\max_{ij}(w_{ij}^5(k))\right|\left\|\max_{jq}(w_{jq}^3(k))\right\|\max_p y_{cp}(k)\right|$$

$$(i = 1,\ldots,m : j = 1,\ldots,n : q = 1,\ldots,r) \qquad (35)$$

According to the definition of the Euclidean norm, we have

$$\left\|\frac{\partial y_i(k)}{\partial W^2}\right\| \leq \sqrt{\frac{mn}{4}}\left|\max_{ij}(w_{ij}^5(k))\right|\left\|\max_{jq}(w_{jq}^3(k))\right\|\max_p y_{cp}(k)\right| \qquad (36)$$

then

$$0 < \eta_2(k) < \frac{8}{mn\left|\max_{ij}(w_{ij}^5(k))\right|\left\|\max_{jq}(w_{jq}^3(k))\right\|\max_p y_{cp}(k)\right|} \qquad (37)$$

We have $\beta_i^2(k) > 0$, then from Eq. (33) we obtain $\Delta E(k) < 0$. According to the Lyapunov stability theory, this shows that the training error will converges to zero as $t \to \infty$. This completes the proof.

Theorem 3 *The stable convergence of the update rule (14) on W^3 is guaranteed if the learning rate $\eta_3(k)$ satisfies the following condition:*

$$0 < \eta_3(k) < \frac{32}{mn\left|\max_{ij}(w_{ij}^5(k))\right|\left\|\max_p y_{c,p}(k)\right|^2} \qquad (38)$$

Proof The error during the learning process can be expressed as

$$e_i(k+1) = e_i(k) + \sum_{j=1}^{n}\sum_{p=1}^{m}\frac{\partial e_i(k)}{\partial w_{jp}^3}\Delta w_{jp}^3 = e_i(k) - \sum_{j=1}^{n}\sum_{p=1}^{m}\frac{\partial y_i(k)}{\partial w_{jp}^3}\Delta w_{jp}^3 \qquad (39)$$

therefore

$$\Delta E(k) = \frac{1}{2}\sum_{i=1}^{m}e_i^2(k)\left[\left(1 - \eta_3(k)[\frac{\partial y_i(k)}{\partial w_{jp}^3}]^T[\frac{\partial y_i(k)}{\partial w_{jp}^3}]^2\right) - 1\right]$$

$$= \frac{1}{2}\sum_{i=1}^{m}e_i^2(k)\left[\left(1 - \eta_3(k)\left\|\frac{\partial y_i(k)}{\partial w_{jp}^3}\right\|^2\right)^2 - 1\right] = -\sum_{i=1}^{m}e_i^2(k)\beta_i^3(k) \qquad (40)$$

where

$$\beta_i^3(k) = \frac{1}{2}\left[1 - \left(1 - \eta_3(k)\left\|\frac{\partial y_i(k)}{\partial w_{jp}^3}\right\|^2\right)^2\right]$$

(41)

therefore

$$\left|\frac{\partial y_i(k)}{\partial w_{jp}^3}\right| = \left|\frac{\partial y_i(k)}{\partial x_j(k)}\cdot\frac{\partial x_j(k)}{\partial w_{jp}^3}\right| = \left|w_{ij}^5 f_j'(.)y_{c,p}(k)\right|$$

$$(i = 1,\ldots,m : j = 1,\ldots,n : p = 1,\ldots,m)$$

(42)

We have

$$\left\|\frac{\partial y_i(k)}{\partial W^3}\right\| < \frac{\sqrt{mn}\left|\max_{ij}(w_{ij}^5(k))\right|\left\|\max_p y_{c,p}(k)\right\|}{4}$$

(43)

then

$$0 < \eta_3(k) < \frac{32}{mn\left|\max_{ij}(w_{ij}^5(k))\right|\left\|\max_p y_{c,p}(k)\right\|^2}$$

(44)

therefore, $\eta_3(k)$ is chosen as above, then we have $\beta_i^3(k) > 0$ and $\Delta E(k) < 0$, According to the Lyapunov stability theory, this shows that the training error will converges to zero as $t \to \infty$.

Theorem 4 *The stable convergence of the update rule (13) on* W^4 *is guaranteed if the learning rate* $\eta_4(k)$ *satisfies the following condition:*

$$0 < \eta_4(k) < \frac{2}{n}$$

(45)

Proof The error during the learning process can be expressed as

$$e_i(k+1) = e_i(k) + \sum_{j=1}^n \frac{\partial e_i(k)}{\partial w_{ij}^4}\Delta w_{ij}^4 = e_i(k) - \sum_{j=1}^n \frac{\partial y_i(k)}{\partial w_{ij}^4}\Delta w_{ij}^4$$

(46)

therefore

$$\Delta E(k) = \frac{1}{2}\sum_{i=1}^m e_i^2(k)\left[\left(1 - \eta_4(k)[\frac{\partial y_i(k)}{\partial W^4}]^T[\frac{\partial y_i(k)}{\partial W^4}]^2\right) - 1\right]$$

$$= \frac{1}{2}\sum_{i=1}^m e_i^2(k)\left[\left(1 - \eta_4(k)\left\|\frac{\partial y_i(k)}{\partial W^4}\right\|^2\right)^2 - 1\right] = -\sum_{i=1}^m e_i^2(k)\beta_i^4(k)$$

(47)

where

$$\beta_i^4(k) = \frac{1}{2}\left[1 - \left(1 - \eta_4(k)\left\|\frac{\partial y_i(k)}{\partial W^4}\right\|^2\right)^2\right] \qquad (48)$$

W^4 represents an n dimensional vector and $\|.\|$ denotes the Euclidean norm. Noticing that the activation function of the hidden neurons in the modified Elman Neural Network is the sigmoidal type, we have

$$\left|\frac{\partial y_i(k)}{\partial w_{il}^4}\right| = \left|x_{c,k}(k)\right| \qquad (i = 1, \ldots, m : l = 1, \ldots, m) \qquad (49)$$

According to the definition of the Euclidean norm, we have

$$\left\|\frac{\partial y_i(k)}{\partial W^4}\right\| < \sqrt{n}\left|x_{c,k}(k)\right| \qquad (50)$$

then

$$0 < \eta_4(k) < \frac{2}{n\left|x_{c,k}(k)\right|^2} \qquad (51)$$

therefore, $\eta_4(k)$ is chosen as:

$$0 < \eta_4(k) < \frac{2}{n\left|x_{c,k}(k)\right|^2} \qquad (52)$$

then we have $\beta_i^4(k) > 0$ and $\Delta E(k) < 0$, according to the Lyapunov stability theory, this shows that the training error will converges to zero as $t \to \infty$.

Theorem 5 *The stable convergence of the update rule (12) on W^5 is guaranteed if the learning rate $\eta_5(k)$ satisfies the following condition:*

$$0 < \eta_5(k) < \frac{2}{mn} \qquad (53)$$

Proof

$$e_i(k+1) = e_i(k) + \sum_{j=1}^{n}\frac{\partial e_i(k)}{\partial w_{ij}^5}\Delta w_{ij}^5 = e_i(k) - \sum_{j=1}^{n}\frac{\partial y_i(k)}{\partial w_{ij}^5}\Delta w_{ij}^5 \qquad (54)$$

therefore

$$\Delta E(k) = \frac{1}{2} \sum_{i=1}^{m} e_i^2(k) \left[\left(1 - \eta_5(k) [\frac{\partial y_i(k)}{\partial W^5}]^T [\frac{\partial y_i(k)}{\partial W^5}]^2 \right) - 1 \right]$$

$$= \frac{1}{2} \sum_{i=1}^{m} e_i^2(k) \left[\left(1 - \eta_5(k) \left\| \frac{\partial y_i(k)}{\partial W^5} \right\|^2 \right)^2 - 1 \right] = - \sum_{i=1}^{m} e_i^2(k) \beta_i^5(k)$$

$$(55)$$

where

$$\beta_i^5(k) = \frac{1}{2} \left[1 - \left(1 - \eta_5(k) \left\| \frac{\partial y_i(k)}{\partial W^5} \right\|^2 \right)^2 \right] \tag{56}$$

W^5 represents an n dimensional vector and $\|.\|$ denotes the Euclidean norm. Noticing that the activation function of the hidden neurons in the modified Elman neural network is the sigmoidal type, we have

$$\left| \frac{\partial y_i(k)}{\partial w_{ij}^5} \right| = |x_j(k)| < 1 \quad (i = 1, \ldots, m : l = 1, \ldots, m) \tag{57}$$

According to the definition of the Euclidean norm, we have

$$\left\| \frac{\partial y_i(k)}{\partial W^5} \right\| < \sqrt{mn} \tag{58}$$

then

$$0 < \eta_4(k) < \frac{2}{mn} \tag{59}$$

therefore, $\eta_5(k)$ is chosen as $0 < \eta_5(k) < (2/mn)$, then we have $\beta_i^5(k) > 0$ and $\Delta E(k) < 0$, According to the Lyapunov stability theory, this shows that the training error will converges to zero as $t \to \infty$.

4 Simulation Results

The objective of this section is to illustrate the performance and capabilities of the proposed structure shown in Fig. 1 for identification of four classes of nonlinear systems considered in [18]. The reference input u (t) to all the identifiers must be selected to be "persistently exciting". For identification of linear systems the persistent excitation of the input guarantees the convergence of the identifier parameters to their true values [1]. The following results are compared with OHF and OIF Elman Neural Network [11] and the amplitude and the frequency of the reference inputs are selected experimentally as recommended in [18].

4.1 Application to Model I

Example 1 The governing equation of the system is given by

$$y(t) = 0.3y(t-1) + 0.6y(t-2) + \frac{0.6}{1 + u(t-1)^2} \qquad (60)$$

where the output at time t is a linear function of past output at times $t-1$ and $t-2$ plus a nonlinear function of the input at time $t-1$. The reference input $u(t-1)$ to the system is selected as $u(t-1) = \sin(2\pi(t-1)/100)$.

To show the robustness of the proposed structure to the variations in the amplitude and frequency of the input, an input with 50 % reduction in the frequency (within the 400–600 time steps) and 100 % increase in the frequency (within the 600–800 time steps) is applied to the system. Figures 2 and 3 depict the simulation results using the OIFHO, OHF and OIF Elman NN.

As can be seen from Fig. 3 the performance of the OIFHO Elman NN structure is more robust to the variations in the amplitude as well as the frequency of the input than two other Networks.

Figure 4 shows the variation of learning rates during the simulation for Example 1. For each step the learning rates are chosen according to Eqs. (30), (37), (44), (52) and (59), in which *max (w)* is chosen from available information for the same step. Selecting learning rates in the determined bounds assures the stability of OIFHO ENN.

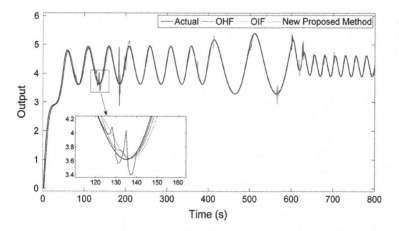

Fig. 2 Responses of the OIFHO, OHF and OIF Elman neural network applied to Example 1 for changing input characteristics

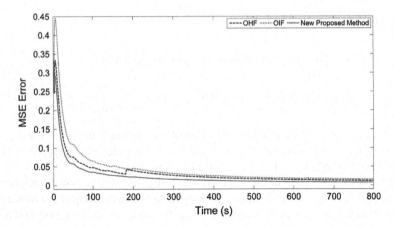

Fig. 3 Comparison of the MSE error of the OIFHO, OHF and OIF Elman neural network applied to Example 1 for changing input characteristics

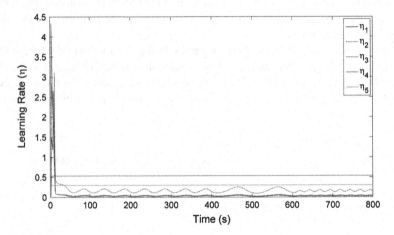

Fig. 4 Variations of learning rates for Example 1

4.2 Application to Model II

Example 2 The governing equation of the system is given by

$$y(t) = \frac{y(t-1)\,y(t-2) + (y(t-2) + 2.5)}{1 + y(t-1)^2 + y(t-2)^2} + u(t-1) \tag{61}$$

where the output at time t is a nonlinear function of the outputs at times $t-1$ and $t-2$ plus a linear function of the input at time $t-1$. The reference input $u(t-1)$ is selected as $u(t-1) = \sin(2\pi(t-1)/25))$.

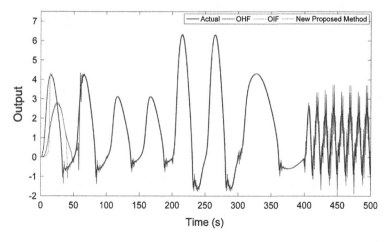

Fig. 5 Responses of the OIFHO, OHF and OIF Elman neural network applied to Example 2 for changing input characteristics

To show the robustness of the proposed structure to variations in the input amplitude and frequency, an input with 50 % reduction in the amplitude (within the 100–200 time steps), 100 % increase in the amplitude (within the 200–300 time steps), 50 % reduction in the frequency (within the 300–400 time steps), and 100 % increase in the frequency (within the 400–500 time steps) is applied to the system. Figures 5 and 6 depict the simulation results using the OIFHO, OHF and OIF Elman NN.

As can be seen from Fig. 6 the performance of the OIFHO Elman NN structure is more robust to the variations in the amplitude as well as the frequency of the input than two other Networks.

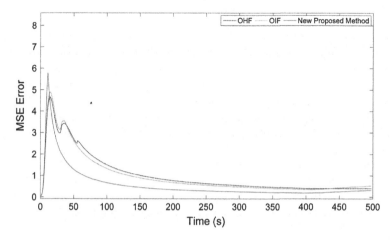

Fig. 6 Comparison of the MSE error of the OIFHO, OHF and OIF Elman neural network applied to Example 2 for changing

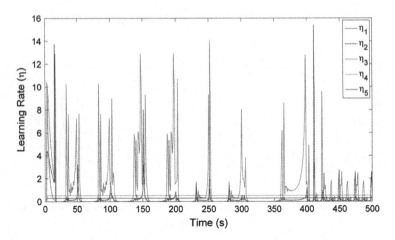

Fig. 7 Variations of learning rates for Example 2

Figure 7 shows the variation of learning rates during the simulation for Example 2. For each step the learning rates are chosen according to Eqs. (30), (37), (44), (52) and (59).

4.3 Application to Model III

Example 3 The governing equation of the system is given by

$$y(t) = \frac{0.2y(t-1) + 0.6y(t-2)}{1 + y(t-1)^2} + \sin(u(t-1)) \tag{62}$$

where the output at time t is a nonlinear function of the output at time $t-1$ and $t-2$ plus a nonlinear function of the input at time $t-1$. The reference input applied to the system is $u(t-1) = \sin(2\pi(t-1)/10)) + \sin(2\pi(t-1)/25))$. To show the robustness of the proposed structure to variations in the input amplitude and frequency, an input with 50 % reduction in the amplitude (within the 400–600 time steps), and 100 % increase in the frequency (within the 600–800 time steps) is applied to the system. Figures 8 and 9 depict the simulation results using the OIFHO, OHF and OIF Elman NN.

As can be seen from Fig. 9 the performance of the OIFHO Elman NN structure is more robust to the variations in the amplitude as well as the frequency of the input than two other Networks. Figure 10 shows the variation of learning rates during the simulation for Example 3. For each step the learning rates are chosen according to Eqs. (30), (37), (44), (52) and (59).

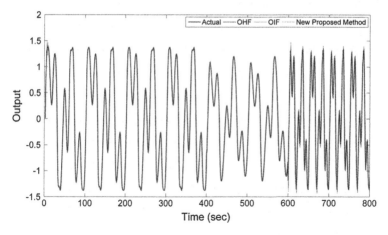

Fig. 8 Responses of the OIFHO, OHF and OIF Elman neural network applied to Example 3 for changing input characteristics

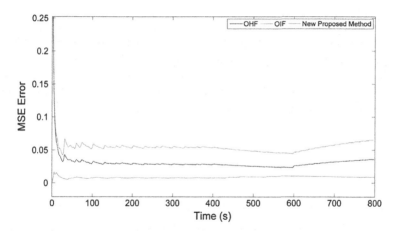

Fig. 9 Comparison of the MSE error of the OIFHO, OHF and OIF Elman neural network applied to Example 3 for changing input characteristics

4.4 Application to Model IV

Example 4 The governing equation of the system is given by

$$y(t) = \frac{y(t-1) + u(t-1)}{1 + y(t-1)^2} \qquad (63)$$

where the output at time t is a nonlinear function of the outputs at times $t - 1$ and the inputs at times $t - 1$. The reference input is $u(t-1) = \sin(2\pi(t-1)/50))$.

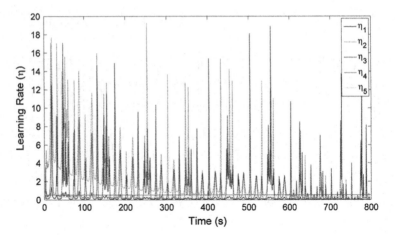

Fig. 10 Variations of learning rates for Example 3

To show the robustness of the proposed structure to variations in the input ampli-tude and frequency, an input with 50 % reduction in the amplitude (within the 250–500 time steps), 100 % increase in the amplitude (within the 500–750 time steps), 50 % reduction in the frequency (within the 750–1,000 time steps), and 100 % increase in the frequency (within the 1,000–1250 time steps) is applied to the neuro-dynamic structure. Figures 11 and 12 depict the simulation results using the OIFHO, OHF and OIF Elman NN.

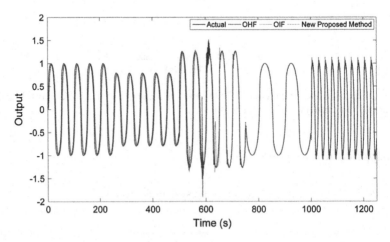

Fig. 11 Responses of the OIFHO, OHF and OIF Elman neural network applied to Example 4 for changing input characteristics

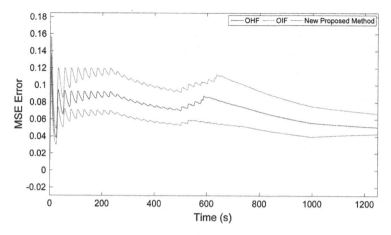

Fig. 12 Comparison of the MSE error of the OIFHO, OHF and OIF Elman neural network applied to Example 4 for changing input characteristics

As can be seen from Fig. 12 the performance of the OIFHO Elman NN structure is more robust to the variations of the amplitude as well as the frequency of the input than two other Networks.

Figure 13 shows the variation of learning rates during the simulation for Example 4. For each step the learning rates are chosen according to Eqs. (30), (37), (44), (52) and (59).

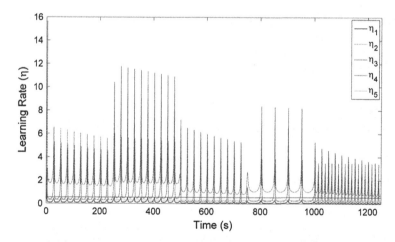

Fig. 13 Variations of learning rates for Example 4

Table 1 Comparison of errors for OIFHO, OHF and OIF structures

	Error	OIFHO	OHF	OIF
Example 1	MSE	0.0078	0.0121	0.0157
	RMSE	0.0882	0.1102	0.1254
	NMSE	0.018	0.0281	0.0365
Example 2	MSE	0.322	0.4073	0.5264
	RMSE	0.5675	0.6382	0.7256
	NMSE	0.0854	0.108	0.1396
Example 3	MSE	0.0312	0.0359	0.0649
	RMSE	0.1765	0.1895	0.2548
	NMSE	0.0405	0.0466	0.0843
Example 4	MSE	0.0427	0.0508	0.0667
	RMSE	0.2067	0.2254	0.2583
	NMSE	0.0603	0.0717	0.0941

5 The Identification Error

Depending on the nature and desired specifications of an application, different error norms may be used to evaluate the performance of an algorithm. We use the Mean Square Error (MSE), Root Mean Square Error (RMSE) and Normalized Mean Square Error (NMSE) to evaluate the performance of OIFHO Elman Neural Network structure proposed in Sect. 2 in comparison with OHF and the OIF structure. The results are given in Table 1.

According to the above tables, we can draw the conclusion that OIFHO structure provided a better performance than other structures, which is due to the excellent nonlinear function approximation capability of OIFHO structure.

6 Conclusion

This paper proposes an improved Elman Neural Network with better performance in comparison with other improved Elman Neural Network by employing three context layers. Subsequently, the dynamic recurrent Backpropagation algorithm for OIFHO is developed according to the gradient descent method. To guarantee the fast convergence, the optimal adaptive learning rates are also derived in the sense of discrete-type Lyapunov stability. Furthermore, capabilities of the proposed structures for identification of four classes of nonlinear systems are shown analytically. Simulation results indicate that the proposed structure is very effective identifying the input–output maps of different classes of nonlinear systems.

References

1. Yazdizadeh, A., Khorasani, K.: Adaptive time delay neural network structure for nonlinear system identification. Neurocomputing **47**, 207–240 (2001)
2. Ge, H.W., Du, W.L., Qian, F., Liang, Y.C.: Identification and control of nonlinear systems by a time-delay recurrent neural network. Neurocomputing **72**, 2857–2864 (2009)
3. Yazdizadeh, A., Khorasani, K.: Identification of a class of nonlinear systems using dynamic neural networks. In: Proceedings of the International Conference on Neural Networks, pp. 161–166. Houston, Texas (1997)
4. Pham, D.T., Liu, X.: Identification of linear and nonlinear dynamic systems using recurrent neural networks. Artif. Intell. Eng. **8**, 67–75 (1993)
5. Wang, J., Zhang, W., Li, Y., Wang, J., Dang, Z.: Forecasting wind speed using empirical mode decomposition and Elman neural network. In: Original Research Article, Applied Soft Computing, Uncorrected Proof, Available online 27 June (2014 in press)
6. Qi. W.M., Chu. C.Y., Ling, J.Q., You, C.W.: A new Elman neural network and its application in hydro-turbine governing system. In: Power and Energy Engineering Conference, APPEEC (2009)
7. Lin, W.M., Hong, C.M.: A new Elman neural network-based control algorithm for adjustable-pitch variable-speed wind-energy conversion systems. IEEE Trans. Power Electron. **26**, 473–481 (2011)
8. Pham, D.T., Liu, X.: Dynamic system modeling using partially recurrent neural networks. J. Syst. Eng. **2**, 90–97 (1992)
9. Hertz, J., Korgh, A., Palmer, R.G.: Recurrent neural networks. Introduction to the Theory of Neural Computing. Addison- wisely, California (1991). Chapter 7
10. Kamijo, K., Tanigawa, T.: Stock price pattern recognition-a recurrent neural network approach. In: Proceedings of the International Joint Conference on Neural Networks, pp. 215–221 (1990)
11. Shi, X.H., Liang, Y.C., Lee, H.P., Lin, W.Z., Xu, X., Lim, S.P.: Improved Elman networks and applications for controlling ultrasonic motors. Appl Artif Intell **18**(7), 603–629 (2004)
12. Ardalani-Farsa, M., Zolfaghari, S.: Chaotic time series prediction with residual analysis method using hybrid Elman-NARX neural networks. Neurocomputing **73**, 2540–2553 (2010)
13. Toqeer, R.S., Bayindir, N.S.: Speed estimation of an induction motor using Elman neural network. Neurocomputing **55**, 727–730 (2003)
14. Seker, S., Ayaz, E., Urkcan, E.T.: Elman's recurrent neural network applications to condition monitoring in nuclear power plant and rotating machinery. Eng. Appl. Artif. Intell. **16**, 647–656 (2003)
15. Song, Q.: On the weight convergence of Elman networks. IEEE Trans. Neural Netw. **21**, 463–480 (2010)
16. Hsu, Chun-Fei: Adaptive backstepping Elman-based neural control for unknown nonlinear systems. Original Research Article Neurocomputing **136**, 170–179 (2014)
17. Cheng, Y.C., Qi, W.M., Zhao, J.: A new Elman neural network and its dynamic properties.In: Cybernetics and Intelligent Systems IEEE Conference, pp. 971–975 (2008)
18. Narendra, K.S., Parthasarathy, K.: Identification and control of dynamical systems using neural networks. IEEE Trans Neural Netw. **1**, 4–27 (1990)

A Simple Model for Evaluating Medical Treatment Options

Irosh Fernando, Frans Henskens, Masoud Talebian and Martin Cohen

Abstract One of the key areas of clinical decision making in the field of clinical medicine involves choosing the most appropriate treatment option for a given patient, out of many alternative treatment options. This paper introduces a model that is intuitive to clinicians for evaluating medication treatment options, and therefore has the advantage of engaging clinicians actively in a collaborative development of clinical Decision Support Systems (DSS). This paper also extends the previously introduced models of medical diagnostic reasoning, and case formulation (in psychiatry). Whilst the proposed model is already implemented as a DSS in psychiatry, it can also be applied in other branches of clinical medicine.

Keywords Model for selecting treatment options · Medical decision support · Medical decision support system

1 Introduction

Clinical reasoning in Medicine can be described in relation to four main areas: diagnostic reasoning, case formulation, choosing investigations, and choosing treatment options. Whilst the authors have previously described a theoretical framework for

I. Fernando (✉) · F. Henskens
School of Electrical Engineering and Computer Science,
University of Newcastle, Callaghan, NSW 2308, Australia
e-mail: irosh.fernando@uon.edu.au

F. Henskens
e-mail: frans.henskens@newcastle.edu.au

M. Talebian
School of Mathematical and Physical Sciences,
University of Newcastle, Callaghan, NSW 2308, Australia
e-mail: masoud.talebian@newcastle.edu.au

M. Cohen
The Mater Hospital, Hunter New England Area Health Service,
Waratah, NSW 2298, Australia
e-mail: martin.cohen@hnehealth.nsw.gov.au

© Springer International Publishing Switzerland 2015
R. Lee (ed.), *Computer and Information Science*, Studies in Computational
Intelligence 566, DOI 10.1007/978-3-319-10509-3_14

diagnostic reasoning, and case formulation [4, 5], this paper mainly focuses on the process of selecting treatment options.

For any given clinical situation, often there are a number of potential treatment options, which are associated with different pros and cons. The process of choosing 'the best' option is typically guided by the clinician's knowledge of the diagnosis and case formulation. Choosing the best option is often a complex process that requires careful evaluation of a number of variables related to each treatment option and the patient's characteristics. Understandably, it is a critical decision that determines the recovery, and considers the risks of potential complications associated with each treatment option. Because of the limitation of human cognitive capacity to process a large number of variables accurately and efficiently in a timely manner, the choice process sometimes results in poor or even adverse outcomes. Therefore, having a theoretical framework for evaluating treatment options in an explicit manner can improve the quality of clinical decision making, and yield benefits to patients.

The first part of the paper explores the process of treatment evaluation at a conceptual level. The next section describes the formalisation of the proposed conceptual model. An example is used to explain the model, and two alternative approaches, namely Analytic Hierarchy Process (AHP) [8], and Genetic Algorithms [6] are briefly compared with the proposed approach. Finally, the paper briefly introduces Treatment Evaluation System (TES), which is an implementation of this model for evaluating treatment options in Psychiatry.

2 Conceptual Model for Evaluating Treatment Options

In order to develop a formal model for evaluating treatment options, it is important to have a conceptual understanding of this process. Gaining such understanding can often be difficult due to the largely implicit nature of clinical reasoning by expert clinicians, and also the domain expertise required in order to conceive the decision making process. The general model often used in modern clinical medicine involves a shared decision making process involving both the clinician and the patient [1]. In this process, the clinician may propose a number of treatment options according to his/her understanding of the diagnosis and etiological formulation, whereas the patient makes an informed decision by evaluating pros and cons associated with each treatment option.

For a given diagnosis, there may be several etiological explanatory models that attempt to explain 'why this patient developed this illness at this point of time?'. Each explanatory model may indicate at least a one treatment option, and collectively there can be a potentially large number of options, out of which a small number of options have to be chosen. As illustrated in Fig. 1, the clinician may look at a large number of variables according to the type of treatment option, and these variables have to be matched against the characteristics of the patient. For example, any given side effect associated with a medication is a one variable, and the matching of this variable with the patient characteristics involves evaluating the risk of this side effect occurring in the patient, and its potential consequences for the patient.

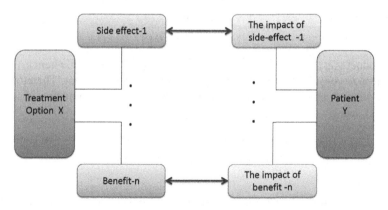

Fig. 1 Conceptual model of treatment evaluation

3 Formal Model

All possible diagnoses consists of a finite set $D = \{d_1, d_2, \ldots\}$, and for any given diagnosis d_i, there exists a set of etiological explanatory models $M(d_i) = \{m_1, m_2, \ldots\}$. For any given explanatory model m_j, there exists a finite set of treatment options $T(m_j) = \{t_1, t_2, \ldots\}$. For each treatment option t_k there exists a finite set of profile items $E(t_k) = \{e_1, e_2, \ldots\}$.

Associated with each treatment option, t_i there exists a row vector of n dimensions $V(t_k) = (v(e_{k1}), \ldots v(e_{kn}))$ where v is a function defined as follows:

$$v : E(t_k) \rightarrow [0 \ldots 1]$$

The reason for this is that the likelihood of the occurrence of any outcome, desirable or undesirable, can be described in terms of their probabilities (e.g. probability of having a successful surgical outcome, probability of having a particular side effect etc.). For example, consider $v(e_{kj}) = 0.6$ representing the probability of the occurrence of the outcome associated with the profile item e_{kj} of the treatment t_k. On the other hand, profile items that are not associated with the probabilities of occurrence (e.g. cost associated with a treatment option) can be assigned a ratio with respect to the largest possible value. For example, consider the profile item e_1 as the treatment cost, and that there are three treatment options: t_1 costs \$1, t_2 costs \$2 and t_3 costs \$3. Then the value of the profile item for each treatment option can be calculated as follows:

$$v(e_{11}) = \frac{\$1}{(\$1 + \$2 + \$3)} \text{ for } t_1$$

$$v(e_{21}) = \frac{\$2}{(\$1 + \$2 + \$3)} \text{ for } t_2$$

$$v(e_{31}) = \frac{\$_3}{(\$_1 + \$_2 + \$_3)} \text{ for } t_3$$

The product $v(e_{kj}).p_j$ can be interpreted as the impact of the profile item e_{kj} of the treatment t_k considering its level of significance p_j to the patient. Therefore, the overall 'fitness' of the treatment t_k can be approximated as a summation of all such products, using a fitness function defined as:

$$f(t_k) = \sum_{j=1}^{n} v(e_{kj}).p_j$$

Suppose there are m of such treatment options, from which at least one needs to be chosen. Collection of the corresponding row vectors associated with these treatment options can be represented as a matrix:

$$M = \begin{pmatrix} v(e_{11}) & \cdots & v(e_{1n}) \\ \vdots & \ddots & \vdots \\ v(e_{m1}) & \cdots & v(e_{mn}) \end{pmatrix}$$

A patient profile corresponding to a treatment option with a profile row vector of n dimensions, can be represented as a column vector of n dimensions:

$$P = \begin{pmatrix} p_1 \\ \vdots \\ p_n \end{pmatrix}$$

where $p_1, \ldots p_n$ represents the relative importance assigned to the profile items $e_{i1}, \ldots e_{in}$ associated with the treatment option t_i. Each p_j is an integer value in the interval $[-9 \ldots 9]$. This is because each treatment option is associated with only two categories of profile item: desirables and undesirables. Negative values correspond to the magnitude of the significance associated with undesirable characteristics of the patient profile (e.g. side effects and adverse complications) whereas positive values correspond to the magnitude of the significance associated with desirable characteristics of the profile (e.g. desirable treatment outcomes). The scales shown in Figs. 2 and 3 can be used to choose a value for desirable and undesirable profile items respectively. The positive scale is somewhat similar to the fundamental scale of absolute numbers used in the Analytic Hierarchy Process (AHP) [8].

Evaluation of the set of m treatment options represented by the $m \times n$ matrix M against the patient profile vector P involves multiplication of M by P resulting in the column vector O of m dimensions, as follows:

$$MP = O$$

Level of importance associated with achieving a desirable outcome	Score
Not important	0
Slight importance	1
Moderate importance	2
Moderate plus importance	3
Strong importance	4
Strong plus importance	5
Very Strong importance	6
Extreme importance	7
Must	8
Absolute Must	9

Fig. 2 Scale for scoring desirable profile items

Level of importance associated with avoiding an undesirable outcome	Score
Not important	0
Slight importance	-1
Moderate importance	-2
Moderate plus importance	-3
Strong importance	-4
Strong plus importance	-5
Very Strong importance	-6
Extreme importance	-7
Must	-8
Absolute Must	-9

Fig. 3 Scale for scoring undesirable profile items

$$\begin{pmatrix} v(e_{11}) & \cdots & v(e_{1n}) \\ \vdots & \ddots & \vdots \\ v(e_{m1}) & \cdots & v(e_{mn}) \end{pmatrix} \begin{pmatrix} p_1 \\ \vdots \\ p_n \end{pmatrix} = \begin{pmatrix} o_1 \\ \vdots \\ o_m \end{pmatrix}$$

The outcome vector O consists of elements representing the relative utility of each treatment option.

4 An Example

In order to explain the model let us consider the following example. Suppose there are three treatment options available for a patient who has a particular diagnosis. Each treatment is associated with a profile vector consisting of five items: probabilities of

Profile items	Treatment 1	Treatment 2	Treatment 3
Probability of having side effect 1	0.1	0.4	0.6
Probability of having side effect 2	0.4	0.5	0.2
Relative cost	0.2	0.4	0.4
Probability of achieving the desirable outcome 1	0.6	0.3	0.9
Probability of achieving the desirable outcome 2	0.7	0.6	0.8

Fig. 4 An example of a matrix of treatment profiles

having each of two side effects, relative cost, and the probabilities of achieving each of two desirable outcomes as described in Fig. 4.

The matrix M corresponding to this table is given as follows:

$$M = \begin{pmatrix} 0.1 & 0.4 & 0.2 & 0.6 & 0.7 \\ 0.4 & 0.5 & 0.4 & 0.3 & 0.6 \\ 0.6 & 0.2 & 0.4 & 0.9 & 0.8 \end{pmatrix}$$

Now, consider the patient profile outlined in Fig. 5. The column vector corresponding to the patient profile is given as:

$$P = \begin{pmatrix} -8 \\ -5 \\ -2 \\ 9 \\ 8 \end{pmatrix}$$

Fig. 5 An example of a patient profile

Profile items	Patient X
Relative impact having the side effect 1	-8
Relative impact of having the side effect 2	-5
Impact of the associated cost	-2
Relative impact of achieving the desirable outcome 1	9
Relative impact of achieving the desirable outcome 2	8

Evaluation of the three treatment options involves the following calculation:

$$MP = O$$

$$\begin{pmatrix} 0.1 & 0.4 & 0.2 & 0.6 & 0.7 \\ 0.4 & 0.5 & 0.4 & 0.3 & 0.6 \\ 0.6 & 0.2 & 0.4 & 0.9 & 0.8 \end{pmatrix} \begin{pmatrix} -8 \\ -5 \\ -2 \\ 9 \\ 8 \end{pmatrix} = \begin{pmatrix} 7.8 \\ 1.0 \\ 7.9 \end{pmatrix}$$

The evaluation outcome vector:

$$O = \begin{pmatrix} 7.8 \\ 1.0 \\ 7.9 \end{pmatrix}$$

represents the relative fitness of the three treatment options. Accordingly, Treatment 3, which has the highest outcome value of 7.9, can be considered the best treatment option.

5 Model Behaviour

Representing the dynamics of the treatment evaluation process as a system of linear equations leads to the advantage that it is more easy to study the behaviour of the system. Understanding of the model's behaviour is necessary for answering some of the important questions in relation to choosing a treatment option.

For example, consider two treatments t_1 and t_2, and that clinician and patient are primarily focused on a particular profile item e_k and its probability of occurrence in this patient in relation to t_1 and t_2. Let us assume that the respective probabilities are $v(e_{1k}) = 0.6$ and $v(e_{2k}) = 0.8$. In this situation the summation of the products $v(e_{1j}).p_j$ and $v(e_{2j}).p_j$ for $i = 1 \ldots n$ and $i \neq k$, are constants; let us assign the values $C_1 = 12$ and $C_2 = 10$ respectively to these sums. This means, without considering the profile item e_k, that treatment t_2 is superior to treatment t_1. One of the useful questions to answer is 'How high a level of significance do you need to assign to profile item e_k so that the treatment t_1 is superior to the treatment t_2'?

The above question can be answered by solving the resulting pair of linear equations:

$$f(t_1) = v(e_{1k})p_k + c_1$$

$$f(t_2) = v(e_{2k})p_k + c_2$$

Setting $f(t_1) = f(t_2)$ and substituting the values for the above problem in these equations gives the following result:

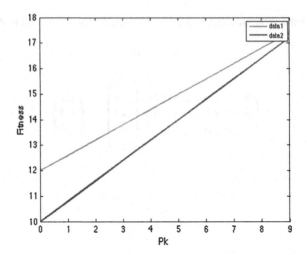

Fig. 6 Behaviour of $f(t_1)$ and $f(t_2)$ according to p_k

$$0.6p_k + 12 = 0.8p_k + 10$$

$$0.2p_k = 2$$

$$p_k = 10$$

Given the above value is out of the range $[-9 \ldots 9]$ the conclusion is that, no matter how important the profile item e_k, treatment t_2 is always superior to treatment t_1. Behaviour of the fitness of each treatment $f(t_1)$ and $f(t_2)$ according to p_k can also be described graphically as shown in Fig. 6.

Now, suppose $v(e_{2k}) = 0.8$ changes to $v(e_{2k}) = 0.85$, whilst $v(e_{1k}) = 0.6$, $C_1 = 12$ and $C_2 = 10$ remain the same (i.e. the level of the evidence base associated with a given treatment property may change slightly over time). Solving the equations for these new values gives:

$$0.6p_k + 12 = 0.85p_k + 10$$

$$0.25p_k = 2$$

$$p_k = 8$$

The above value is within the range $[-9 \ldots 9]$, and according to the scale given in Fig. 2, profile item e_k is a 'must' to achieve. This new result can be interpreted as saying that both treatments have the same degree of fitness, if the patient considers e_k as a 'must' to achieve. Nevertheless, if the patient changes his/her mind and assigns e_k as an 'absolute must' to achieve (i.e. $p_k = 8$), then treatment t_1 is superior to

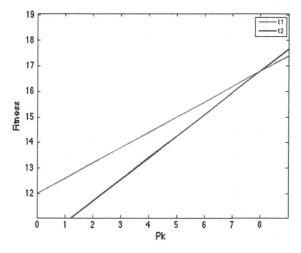

Fig. 7 Behaviour of $f(t_1)$ and $f(t_2)$ according to p_k

treatment t_2, and indeed is the only treatment that can satisfy the 'absolute must' requirement. Figure 7 describes this situation.

The above example can easily be extended to study the behaviour of more than two treatments, and also for more than one profile item. Studying the model's behaviour is important to gaining deeper understanding of the treatment evaluation process. Importantly, the model incorporates the clinician's expertise by supporting him/her in adjusting the values of profile items in order to make fair and effective decisions.

6 Alternative Approaches

It is important to recognise that there are problems in other domains that require similar mathematical models at an abstract level, and therefore there exist alternative strategies that could potentially be applied for evaluating treatment options.

For example, Analytic Hierarchy Process (AHP) is a well-established and mathematically rigorous procedure [8]. AHP has been widely applied in various application domains including clinical practice [7]. For example, application of AHP has been described in relation to optimal management of pharyngitis [9], and also estrogen replacement therapy and cosmetic eye lid surgery [10]. Application of AHP requires a pairwise comparison of profile items, and for n profile items, $n(n - 1)/2$ (i.e. $O(n^2)$) judgments have to be made. The proposed approach only requires n (i.e. $O(n)$) judgments to be made.

For example consider a treatment profile with a vector of four desirable outcomes. Formulation of the problem in terms of AHP requires rating the relative importance of these profile items using a scale of absolute numbers. Suppose the resulting pairwise comparison matrix is:

$$[a_{ij}] = \begin{pmatrix} 1 & 3 & 6 & 7 \\ 1/3 & 1 & 1/5 & 1/4 \\ 1/6 & 5 & 1 & 3 \\ 1/7 & 4 & 1/3 & 1 \end{pmatrix}$$

Then, for this example, the item $a_{12} = 3$ is interpreted as saying that profile item 1 is three times more important than profile item 2.

Whilst comparison of profile items of the same category (e.g. undesirable profile item with another undesirable, or desirable property with another desirable property) is clearly meaningful, in the context of the evaluation of treatment choices, comparison of profile items in different categories (e.g. cost of the treatment with a side effect, or an undesirable property with a desirable property) is difficult and sometimes not meaningful.

On the other hand, the pairwise matrix has to be a positive matrix, and if the matrix of treatment profile vectors with positive and negative values is transformed into a positive matrix its interpretation become less intuitive to clinicians. Also, the AHP algorithm requires many complex calculations (e.g. the principal eigenvector, Perron vector, and their eigenvalues) and therefore requires more computational resources. More importantly, as the authors have previously emphasised, engagement of clinicians in a collaborative development environment is a critical step for successful development of Clinical Decision Support Systems [3]. AHP would be less appealing due to its complexity, and may appear less intuitive to clinicians.

Genetic Algorithms (GA) can also be applied to solve problems of similar nature. In GA, for example, profile items can be encoded as genes with their initial values, and the genetic operations such as cross over and mutations can be applied to produce a pool of profile vectors with different values. A fitness function can be designed to select the 'the fittest' profile.

GA is better suited to situations which require selecting a best solution out of a large number of solutions. For example, consider a hypothetical situation in which a pharmaceutical treatment can be designed by adjusting the doses of different chemical components that are correlated with corresponding profile item values. This may result in a potentially infinite number of possible combinations of different chemical components, and thus an infinite number of treatment profiles. Given n different chemical compounds required to synthesise a treatment, a gene can be encoded as a vector $g_0 = (w_1, \ldots, w_n)$ where w_i is the amount of the ith required chemical component. Using the above-mentioned genetic operations a very large (infinitely many) pool of genes can be replicated. Suppose the functions Ω_i where $i = 1 \ldots n$ determines the corresponding values of the profile items e_i such that:

$$v(e_i) = \Omega_i(w_i)$$

Then the fitness of any given gene g_k can be evaluated using a fitness function f as follows:

$$f(g_k) = \sum_{i=1}^{n} \Omega_i(w_i)$$

The 'best' treatment option can be chosen out of any given set of treatment options that are encoded in genes by choosing the corresponding gene with the highest fitness value.

Evaluation of medical treatment options often involves only a few options, and therefore GA is less desirable.

7 Model Implementation in Psychiatry

The new model described above has been implemented as Treatment Evaluation System (TES) in psychiatry, and used for choosing psychiatric treatment options. The design, implementation and its application is presented elsewhere in a separate paper [2]. Figure 8 shows a screenshot of TES, in which two antidepressant treatments are evaluated against a hypothetical patient profile.

Next, Fig. 9 shows approximated values for each treatment profile.

Finally, Fig. 10 shows the evaluation results, after entering the patient profile.

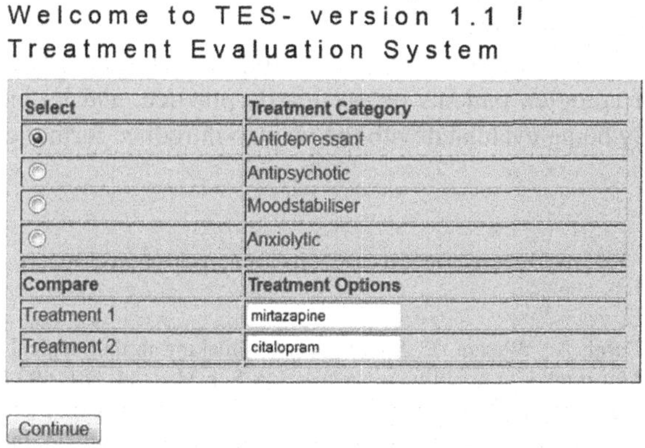

Welcome to TES- version 1.1 !
Treatment Evaluation System

Select	Treatment Category
⊙	Antidepressant
○	Antipsychotic
○	Moodstabiliser
○	Anxiolytic

Compare	Treatment Options
Treatment 1	mirtazapine
Treatment 2	citalopram

Continue

Fig. 8 Implementation of the model as treatment evaluation system in psychiatry

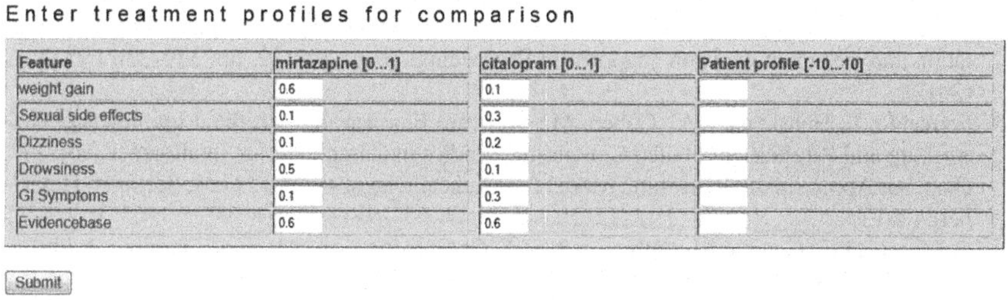

Enter treatment profiles for comparison

Feature	mirtazapine [0...1]	citalopram [0...1]	Patient profile [-10...10]
weight gain	0.6	0.1	
Sexual side effects	0.1	0.3	
Dizziness	0.1	0.2	
Drowsiness	0.5	0.1	
GI Symptoms	0.1	0.3	
Evidencebase	0.6	0.6	

Submit

Fig. 9 Treatment profiles and the patient profile

Treatment Evaluation Results:

mirtazapine	citalopram	Patient
0.6	0.1	-8
0.1	0.3	-7
0.1	0.2	-4
0.5	0.1	5
0.1	0.3	-2
0.6	0.6	9
Value: 1.8	Value 1.5999999	

Close

Fig. 10 Outcome of the treatment evaluation

8 Conclusion

This paper presents a new model that can be used for effectively evaluating competing treatment options. The model has been described at a relatively abstract level, and encapsulates the essence of treatment decision making across different branches of clinical medicine. Therefore, the model can be implemented as a decision support tool in any branch of clinical medicine irrespective of the nature of the involved treatment options. The proposed model was originally formulated to prescribe psychotropic medications for complex patients in psychiatric practice, and its implementation, TES, is currently being evaluated with the view to introduce further enhancements.

References

1. Charles, C., Gafni, A., Whelan, T.: Shared decision-making in the medical encounter: what does it mean? (or it takes at least two to tango). Soc. Sci. Med. **44**, 681–692 (1997)
2. Fernando, I., Henskens, F.A.: A web-based flatform for collaborative development of a knowledgebase for psychiatric case formulation and treatment decision support. In: IADIS e-Health 2012 International Conference. Lisban, Portugal (2012)
3. Fernando, I., Henskens, F.A., Cohen, M.: A collaborative and layered approach (CLAP) for medical expert systems development: a software process model. In: Proceedings, SERA (2012)
4. Fernando, I., Henskens, F.A., Cohen, M.: An expert system model in psychiatry for case formulation and treatment decision support. In: Proceedings, HealthInf-12, pp. 329–336 INSTICC (2012)
5. Fernando, I., Henskens, F.A., Cohen, M.: Software Engineering, Artificial Intelligence, Networking and Parallel/Distributed Computing, Studies in Computational Intelligence, vol. 492, chap. An Approximate Reasoning Model for Medical Diagnosis, pp. 11–24, Springer, Heidelberg (2013)
6. Goldberg, D.E.: Genetic Algorithms in Search, Optimization, and Machine Learning. Addison-Wesley Professional, New York (1989)
7. Liberatore, M.J., Nydick, R.L.: The analytic hierarchy process in medical and health care decision making: a literature review. Eur. J. Oper. Res. **189**, 194–207 (2008)

8. Saaty, T.L.: How to make a decision: the analytic hierarchy process. Eur. J. Oper. Res. **48**, 9–26 (1990)
9. Singh, S., Dolan, J., Centor, R.: Optimal management of adults with pharyngitis - a multi-criteria decision analysis. BMC Med. Inform. Decis. Mak. **6**(1), 14 (2006)
10. Singpurwalla, N., Forman, E., Zalkind, D.: Promoting shared health care decision making using the analytic hierarchy process. Socio-Econ. Plann. Sci. **33**, 277–299 (1999)

Author Index

© Springer International Publishing Switzerland 2015
R. Lee (ed.), *Computer and Information Science*, Studies in Computational
Intelligence 566, DOI 10.1007/978-3-319-10509-3

Printed in the United States
By Bookmasters